Operations Management

Operations Management

Edited by **Edward Pepper**

New York

Published by NY Research Press,
23 West, 55th Street, Suite 816,
New York, NY 10019, USA
www.nyresearchpress.com

Operations Management
Edited by Edward Pepper

International Standard Book Number: 978-1-63238-354-9 (Hardback)

Contents

Preface

This book has been an outcome of determined endeavour from a group of educationists in the field. The primary objective was to involve a broad spectrum of professionals from diverse cultural background involved in the field for developing new researches. The book not only targets students but also scholars pursuing higher research for further enhancement of the theoretical and practical applications of the subject.

Comprehensive information regarding the field of operations management has been encompassed in this profound book. Operations management is a field of business concerned with managing the procedure that transforms inputs into outputs, in the form of goods and/or services. Due to the increase in complex environments along with the current economic swings and substantially squeezed industrial margins, there is extra pressure on corporations, and decision makers are pushed to enhance operational efficiency and effectiveness. This book includes contributions by various experts which demonstrate novel ideas, original results, practical experiences and a systematization of some basic issues in operations management. This book will be useful for readers from various backgrounds interested in this field.

It was an honour to edit such a profound book and also a challenging task to compile and examine all the relevant data for accuracy and originality. I wish to acknowledge the efforts of the contributors for submitting such brilliant and diverse chapters in the field and for endlessly working for the completion of the book. Last, but not the least; I thank my family for being a constant source of support in all my research endeavours.

Editor

Improving Operations Performance with World Class Manufacturing Technique: A Case in Automotive Industry

Fabio De Felice, Antonella Petrillo and
Stanislao Monfreda

Additional information is available at the end of the chapter

1. Introduction

Global competition has caused fundamental changes in the competitive environment of manufacturing industries. Firms must develop strategic objectives which, upon achievement, result in a competitive advantage in the market place. However, for almost all manufacturing industries, an increased productivity and better overall efficiency of the production line are the most important goals. Most industries would like to find the formula for the ultimate productivity improvement strategy. Industries often suffer from the lack of a systematic and consistent methodology. In particular the manufacturing world has faced many changes throughout the years and as a result, the manufacturing industry is constantly evolving in order to stay ahead of competition [1]. Innovation is a necessary process for the continuous changes in order to contribute to the economic growth in the manufacturing industry, especially to compete in the global market. In addition to innovation as a mode for continued growth and change, there are many other vehicles for growth in the manufacturing industry [2], [3]. One in particular that has been gaining momentum is the idea of World Class Manufacturing (WCM) developed by Richard J. Schonberger (in the 80s) who collected several cases, experiences and testimonies of companies that had embarked on the path of continuous "Kaizen" improvement for excellence in production, trying to give a systematic conception to the various practices and methodologies examined. Some of the benefits of integrating WCM include increased competitiveness, development of new and improved technology and innovation, increased flexibility, increased communication between management and production employees, and an increase in work quality and workforce empowerment. This work takes you to the journey of World Class Manufacturing System (WCMS) adopted by the most important automotive Company located in Italy, the Fiat Group Automobiles. World class can be defined as a tool used to search and allow a company to perform at a best-on-class level.

The aim of this work is to present establishments of the basic model of World Class Manufacturing (WCM) quality management for the production system in the automotive industry in order to make products of the highest quality eliminating losses in all the factory fields an improvement of work standards.

The chapter is organized as follows: Section 2 introduces World Class Manufacturing and illustrates literature review, mission and principles of WCM, Section 3 describes Tools for WCM with particular attention on their features and on Key Performance and Key Activities Indicators and Section 4 describes the research methodology through a real case study in the largest Italian automotive company. To conclude, results and conclusions are provided.

2. Literature review

Manufacturers in many industries face worldwide competitive pressures. These manufacturers must provide high-quality products with leading-edge performance capabilities to survive, much less prosper. The automotive industry is no exception. There is intense pressure to produce high-performance at minimum-costs [4]. Companies attempting to adopt WCM have developed a statement of corporate philosophy or mission to which operating objectives are closely tied. A general perception is that when an organization is considered as world-class, it is also considered as the best in the world. But recently, many organizations claim that they are world-class manufacturers. Indeed we can define world class manufacturing as a different production processes and organizational strategies which all have flexibility as their primary concern [5]. For example Womack et al. [6] defined a lead for quantifying world class. Instead Oliver et al. [7] observed that to qualify as world class, a plant had to demonstrate outstanding performance on both productivity and quality measures. Summing up we can state that the term World-Class Manufacturing (WCM) means the pursuance of best practices in manufacturing. On the other hand we would like to note that one of the most important definition is due to Schonberger. He coined the term "World Class Manufacturing" to cover the many techniques and technologies designed to enable a company to match its best competitors [8].

When Schonberger first introduced the concept of "World Class Manufacturing", the term was seen to embrace the techniques and factors as listed in Figure 1. The substantial increase in techniques can be related in part to the growing influence of the manufacturing philosophies and economic success of Japanese manufacturers from the 1960s onwards. What is particularly interesting from a review of the literature is that while there is a degree of overlap in some of the techniques, it is clear that relative to the elements that were seen as constituting WCM in 1986, the term has evolved considerably.

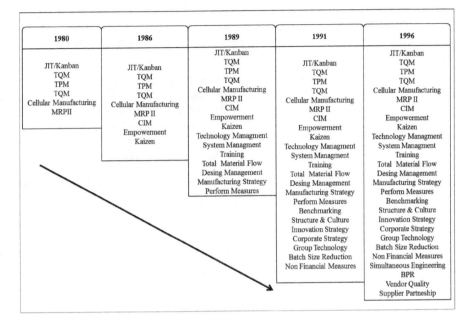

1980	1986	1989	1991	1996
JIT/Kanban TQM TPM TQM Cellular Manufacturing MRPII	JIT/Kanban TQM TPM TQM Cellular Manufacturing MRP II CIM Empowerment Kaizen	JIT/Kanban TQM TPM TQM Cellular Manufacturing MRP II CIM Empowerment Kaizen Technology Managment System Managment Training Total Material Flow Desing Management Manufacturing Strategy Perform Measures	JIT/Kanban TQM TPM TQM Cellular Manufacturing MRP II CIM Empowerment Kaizen Technology Managment System Managment Training Total Material Flow Desing Management Manufacturing Strategy Perform Measures Benchmarking Structure & Culture Innovation Strategy Corporate Strategy Group Technology Batch Size Reduction Non Financial Measures	JIT/Kanban TQM TPM TQM Cellular Manufacturing MRP II CIM Empowerment Kaizen Technology Managment System Managment Training Total Material Flow Desing Management Manufacturing Strategy Perform Measures Benchmarking Structure & Culture Innovation Strategy Corporate Strategy Group Technology Batch Size Reduction Non Financial Measures Simultaneous Engineering BPR Vendor Quality Supplier Partneship

Figure 1. The growth of techniques associated with the WCM concept

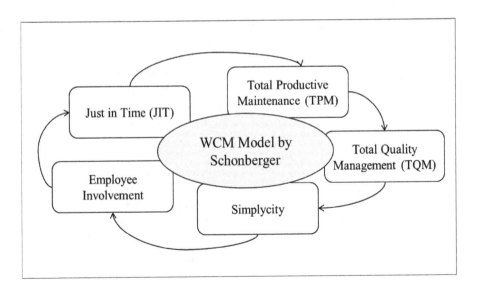

Figure 2. WCM Model by Schonberger

These techniques have been known for a long time, but with Schonberger, a perfectly integrated and flexible system was obtained, capable of achieving company competitiveness with products of high quality. The WCM model by Schonberger is illustrated here above in Figure 2.

According to Fiat Group Automobiles, "World Class Manufacturing (WCM)" is: a structured and integrated production system that encompasses all the processes of the plant, the security environment, from maintenance to logistics and quality. The goal is to continuously improve production performance, seeking a progressive elimination of waste, in order to ensure product quality and maximum flexibility in responding to customer requests, through the involvement and motivation of the people working in the establishment.

The WCM program has been made by Prof. Hajime Yamashina from 2005 at the Fiat Group Automobiles. The program is shown here below in Figure 3.

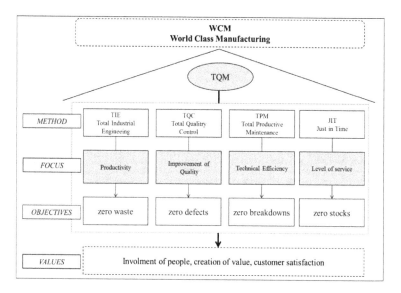

Figure 3. World Class Manufacturing in Fiat Group Automobiles

Fiat Group Automobiles has customized the WCM approach to their needs with Prof. Hajime Yamashina from Kyoto University (he is also member of the Royal Swedish Academy and in particular he is RSA Member of Engineering Sciences), by redesigning and implementing the model through two lines of action: **10 technical pillars**; **10 managerial pillars.**

The definition proposed by Yamashina includes a manufacturing company that excels in applied research, production engineering, improvement capability and detailed shop floor knowledge, and integrates those components into a combined system. In fact, according to Hajime Yamashina the most important thing continues to be the ability to change and quickly [9]. WCM is developed in 7 steps for each pillar and the steps are identified in three phases: *reactive, preventive and proactive*. In figure 4 an example of a typical correlation between steps and phases is shown, but this correlation could change for each different technical pillar; in fact each pillar could have a different relation to these phases. The approach of WCM needs to start from a "**model area**" and then extend to the entire company. WCM "attacks" the manufacturing area. WCM is based on a system of audits that give a score that allows to get to the highest level. The highest level is represented by "*the world class level*".

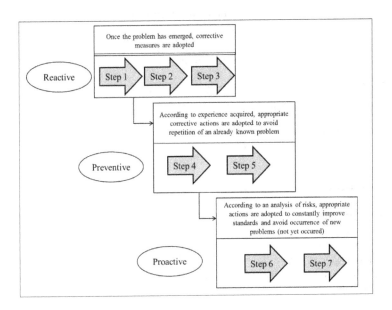

Figure 4. World Class Manufacturing steps

2.1. Mission and principles

The process to achieve "World Class Manufacturing" (WCM) has a number of philosophies and elements that are common for all companies. Therefore, when applied to the manufacturing field, TQM and WCM are synonymous. We would like to observe that customer needs and expectations is a very important element in WCM. The manufacturing strategy should be geared to support these needs. These could be dealing with certification, market share, company growth, profitability or other global targets. The outcomes should be defined so that they are measurable and have a definite timetable. These are also a means of defining employee responsibilities and making them feel involved. Employee education and training is an essential element in a World Class Manufacturing Company. They must understand the company's vision and mission and consequential priorities. As introduced in World Class Manufacturing, well known disciplines such as: Total Quality Control; Total

Productive Maintenance; Total Industrial Engineering; Just In Time and Lean Manufacturing are taken into account. Thus, World Class Manufacturing is based on a few fundamental principles:

- the involvement of people is the key to change;
- it is not just a project, but a new way of working,
- accident prevention is a non-derogated "value";
- the customer's voice should reach all departments and offices;
- all leaders must demand respect for the standards set;
- methods should be applied with consistency and rigor;
- all forms of MUDA waste are not tolerable;
- all faults must be made visible;
- eliminate the cause and not treat the effect.

2.2. Pillars: Description and features

WCM foresees 10 technical pillars and 10 managerial pillars. The levels of accomplishment in technical fields are indirectly affected by the level of accomplishment in administrative fields. The pillar structure represents the "Temple of WCM" (Figure 5) and points out that, to achieve the standard of excellence, a parallel development of all the pillars is necessary. Each pillar focuses on a specific area of the production system using appropriate tools to achieve excellence global.

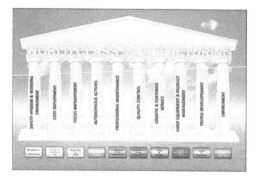

Figure 5. Temple of WCM

Here below in Table 1 features for each technical pillars are illustrated.

Technical Pillar	Why	Purpose
SAF Safety	Continuous improvement of safety	To reduce drastically the number of accidents. To develop a culture of prevention. To improve the ergonomics of the workplace. To develop specific professional skills.
CD Cost Deployment	Analysis of the losses and costs (losses within the costs)	To identify scientifically and systematically the main items of loss in the system production-logistics business. To quantify the potential economic benefits and expected. To address the resources and commitment to managerial tasks with greatest potential.
FI Focused Improvement	Priorities of actions to management the loss identified by the cost deployment	To reduce drastically the most important losses present in the system manufacturing plant, eliminating inefficiencies. To eliminate non-value-added activities, in order to increase the competitiveness of the cost of the product. To develop specific professional skills of problem solving.
AA Autonomous Activities	Continuous improvement of plant and workplace	It is constituted by two pillars: AM Autonomous Maintenance. It is used to improve the overall efficiency of the production system through maintenance policies through the conductors (equipment specialists). WO Workplace Organization. It is develops to determine an improvement in the workplace, because often the materials and equipment are degrade; in particular because in the process there are many losses (MUDA)to remove.
PM Professional Maintenance	Continuous improvement of downtime and failures	To increase the efficiency of the machines using failure analysis techniques. To facilitate the cooperation between conductors (equipment specialists) and maintainers (maintenance people) to reach zero breakdowns.
QC Quality Control	Continuous improvement of customers' needs	To ensure quality products. To reduce non-compliance. To increase the skills of the employees.
LOG Logistics & Customer Service	Optimization of stocks	To reduce significantly the levels of stocks. To minimize the material handling, even with direct deliveries from suppliers to the assembly line.
EEM Early Equipment Management EPM Early Product Management	Optimization of installation time and costs and optimization of features of new products	To put in place new plants as scheduled. To ensure a rapid start-up and stable. To reduce the Life Cycle Cost (LCC). To design systems easily maintained and inspected.

Technical Pillar	Why	Purpose
PD People Development	Continuous improvement of the skills of employees and workers	To ensure, through a structured system of training, correct skills and abilities for each workstation. To develop the roles of maintenance workers, technologists, specialists such as major staff training.
ENV Environment ENE Energy	Continuous improvement environmental management and reduce energy waste	To comply with the requirements and standards of environmental management. To develop an energy culture and to reduce the energy costs and losses.

Table 1. Description of pillars

As regards the ten Managerial Pillars there are: 1) Management Commitment; 2) Clarity of Objectives; 3) Route map to WCM; 4) Allocation of Highly Qualified People to Model Areas; 5) Organization Commitment; 6) Competence of Organization towards Improvement; 7) Time and Budget; 8)Detail Level; 9) Expansion Level and 10) Motivation of Operators

3. The main tools for World Class Manufacturing: features and description

WCM requires all decisions to be made based on objective measured data and its analysis. Therefore, all the traditional data analysis tools such as scatter diagrams, histograms and checklists are used. Thus, from literature survey it is inferred that it is not possible to use the specific single tool to achieve world-class performance and address all the manufacturing components. It is inferred that to address all the components of the manufacturing system the following tools are necessary (see Table 2):

Main Tools	Description
5 G	It is a methodology for the description and the analysis of a loss phenomenon (defects, failures malfunctions...). It based on the facts and the use of the 5 senses
4M or 5M	It is used by the list of possible factors (causes, sub-causes) that give rise to the phenomenon. For the 4M the causes are grouped into 4 categories: Methods; Materials; Machines; Mans. And for the 5M, there are the same 4M more the fifth that is the environment.
5 S	It is used to achieve excellence through improvement of the workplace in terms of order, organization and cleanliness. The technique is based on: Seiri (separate and order); Seiton (arrange and

Main Tools	Description
	organize); Seiso (clean); Seiketsu (standardized); Shitsuke (maintaining and improving).
5W + 1H	It is used to ensure a complete analysis of a problem on all its fundamental aspects. The questions corresponding to the 5 W and 1 H are: Who? What? Why? Where? When? How?
5 Whys	It is used to analyze the causes of a problem through a consecutive series of questions. It is applied in failures analysis, analysis of sporadic anomalies, analysis of chronic losses arising from specific causes.
AM Tag	It is a sheet which, suitably completed, is applied on the machine, in order to report any anomaly detected.
WO Tag	It is a sheet which, suitably completed, is used in order to report any anomaly detected for Workplace Organization
PM Tag	It is a sheet which, suitably completed, is used in order to report any anomaly detected for Professional Maintenance.
Heinrich Pyramid	It is used for classifying the events that have an impact on safety such as fatalities, serious, minor, medications, near-accidents, accidents, dangerous conditions and unsafe practices over time.
SAF Tag	It is a sheet which, suitably completed, is used in order to report any anomaly detected for Safety.
Equipment ABC Prioritization	It is used to classify plants according their priorities of intervention in case of failure.
Cleaning cycles	Are used for activities on Autonomous Maintenance, Workplace Organization and Professional Maintenance.
Inspection cycles	Are used for activities on Autonomous Maintenance, Workplace Organization and Professional Maintenance.
Maintenance cycles	Are used for activities on Autonomous Maintenance and Professional Maintenance.
Control cycles	Are used for activities on Autonomous Maintenance, Workplace Organization and Professional Maintenance.
FMEA-Failure Mode and Effect Analysis	It is used to prevent the potential failure modes.
Kanban	It is a tag used for programming and production scheduling.
Kaizen (Quick, Standard, Major, Advanced)	It is a daily process, the purpose of which goes beyond simple productivity improvement. It is also a process that, when done correctly, humanizes the workplace, eliminates overly hard work.
Two Videocamera Method	It is used to perform the video recording of the transactions in order to optimize them.
MURI Analysis	Ergonomic analysis of workstations.

Main Tools	Description
MURA Analysis	Analysis of irregular operations.
MUDA Analysis	Analysis of losses.
Spaghetti Chart	It is a graphical used to detail the actual physical flow and distances involved in a work process.
Golden Zone & Strike zone Analysis	Analysis of work operations in the area that favors the handling in order to minimize movement to reduce fatigue.
OPL (One Point Lesson)	It is a technique that allows a simple and effective focus in a short time on the object of the training.
SOP (Standard Operation Procedure)	Standard procedure for work.
JES (Job Elementary Sheet)	Sheet of elementary education.
Visual Aid	It is a set of signals that facilitates the work and communication within the company.
Poka Yoke	It is a prevention technique in order to avoid possible human errors in performance of any productive activity.
TWTTP (The way to teach people)	It is an interview in 4 questions to test the level of training on the operation to be performed.
HERCA (Human Error Root Cause Analysis)	It is a technique for the investigation of events of interest, in particular accidents, which examines what happened researching why it happened.
RJA (Reconditional Judgment Action Analysis)	Analysis of judgment, recognition and action phases at work.
5Q 0D (Five Questions to Zero Defects)	Analysis of the process or of the equipment (machine) through five questions to have zero defect.
DOE	It is a techniques enables designers to determine simultaneously the individual and interactive effects of many factors that could affect the output results in any design.
ANOVA	It is a collection of statistical models, and their associated procedures, in which the observed variance in a particular variable is partitioned into components attributable to different sources of variation.
PPA (Processing Point Analysis)	It is used for restore, maintain and improve operational standards of work by ensuring zero defects.
QA Matrix (Matrix Quality Assurance)	It is a set of matrices which shows the correlations between the anomalies of the product and the phases of the production system.
QM Matrix (Matrix Maintenance Quality)	It is a tool used to define and maintain the operating conditions of the machines which ensure performance of the desired quality.
QA Network quality assurance network	It is used to ensure the quality of the process by eliminating rework.

Main Tools	Description
QuOA quality operation analysis	Preventive analysis of the work steps to ensure the quality.
SMED (Single Minute Exchange of Die)	It is a set of techniques to perform operations of development, set-up, with a duration < 10 minutes.
Rhythmic operation analysis	Analysis of the dispersion during the work cycle.
Motion Economic Method	Analysis used to evaluate the efficiency of movement and optimize them.
Value Stream Map	It allows to highlight the waste of a process business, helping to represent the current flow of materials and information that, in relation to a specific product, through the value stream between customer and suppliers.
Material Matrix	Classification of materials according to tree families A – B - C and subgroups.
X Matrix	It is a tool for quality improvement, which allows compare two pairs of lists of items to highlight the correlations between a list, and the two adjacent lists. X matrix to relate defect mode, phenomenon, equipment section and quality components.

Table 2. Main Tools and description

3.1. Key Performance Indices and Key Activity Indicators

In World Class Manufacturing the focus is on continuous improvement. As organizations adopt world class manufacturing, they need new methods of performance measurement to check their continuous improvement. Traditional performance measurement systems are not valid for the measurement of world class manufacturing practices as they are based on outdated traditional cost management systems, lagging metrics, not related to corporate strategy, inflexible, expensive and contradict continuous improvement [10]. To know the world class performance, measurement is important because *"if you can't measure it, you can't manage it and thus you can't improve upon it"*.

Here below in Table 3 is shown a brief report on different indices and indicators defined by several authors in order to "measure" WCM.

However, some authors [15; 16] proposed only productivity as a measure of manufacturing performance. Kennerley and Neely [17] identified the need for a method that could be used for the development of measures able to span diverse industry groups. From this point of view we would like to note that it is necessary to develop a more systematic approach in order to improve a project and process. In particular, in WCM we can use two types of indicators: Key Performance Indicator (KPI) and Key Activity Indicator (KAI). KPI represents a result of project improvement, e.g. sales, profit, labor productivity, equipment performance rate, product quality rate, Mean Time to Failure (MTBF) and Mean Time to Repair (MTTR)

Indices/Indicators	Authors			
	Kodali et al. [11]	Wee and Quazi [12]	Digalwar and Metri [13]	Utzig [14]
Broad management/Worker involvement				+
Competitive advantage	+			
Cost/Price	+		+	+
Customer relations/Service	+		+	
Cycle time				+
Engineering change notices				+
Facility control			+	
Flexibility	+		+	+
Global competitiveness			+	
Green product/Process design		+	+	
Innovation and Technology			+	
Inventory	+		+	
Machine hours per part				+
Measurement and information management.		+		
Morale	+			
Plant/Equipment/Tooling reliability				+
Problem support				+
Productivity	+			+
Quality	+		+	+
Safety	+		+	
Speed/Lead Time			+	+
Supplier management		+		
Top management commitment		+	+	
Total involvement of employees		+		
Training		+		

Table 3. Main indices/indicators defined by different authors

[18, 19]. KAI represents a process to achieve a purpose of project improvement, e.g. a total number of training cycles for employees who tackle performance improvement projects, a total number of employees who pass a public certification examination and an accumulative number of Kaizen cases [20]. A KAI & KPI overview of a Workplace Organization is seen in Figure 6 (Autonomous Activities pillar).

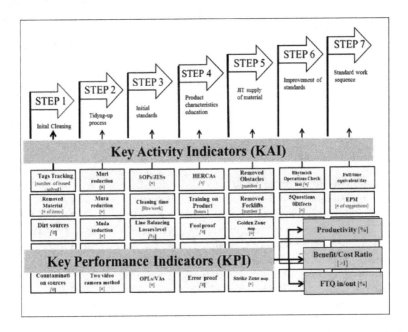

Figure 6. A KAI & KPI overview of a Workplace Organization

4. Industrial case study

The aim of this work is to present establishments of the basic model of World Class Manufacturing (WCM) quality management for the production system at Fiat Group Automobiles in order to make products of the highest quality eliminating losses in all the factory fields an improvement of work standards. In fact, World Class Manufacturing is a manufacturing system defined by 6 International companies including Fiat Group Automobiles with the intent to raise their performances and standards to World Class level with the cooperation of leading European and Japanese experts and this includes all the plant processes including quality, maintenance, cost management and logistics etc. from a universal point of view. Thus, automotive manufacturing requires the ability to manage the product and its associated information across the entire fabricator. Systems must extend beyond their traditional role of product tracking to actively manage the product and its processing. This requires co-ordinating the information flow between process equipment and higher level systems, supporting both manual and automatic interfaces. A case study methodology was used to collect detailed information on division and plant strategic objectives, performance measurement systems, and performance measurement system linkages. The result of this research was to develop principles on strategic objectives, performance measurement systems and performance measurement system linkages for improved organizational coordination. The purpose of this study is to examine the relationship between division and plant performance measurement systems designed to support the firm's strategic objectives and to improve organizational coordination. We will focus our attention on the Cost Deployment Pillar, Autonomous Activities/Workplace Organization Pillar and Logistics/Customers Service Pillar.

4.1. Company background

Fiat Group Automobiles is an automotive-focused industrial group engaged in designing, manufacturing and selling cars for the mass market under the Fiat, Lancia, Alfa Romeo, Fiat Professional and Abarth brands and luxury cars under the Ferrari and Maserati brands. It also operates in the components sector through Magneti Marelli, Teksid and Fiat Powertrain and in the production systems sector through Comau. Fiat operates in Europe, North and South America, and Asia. Its headquarters is in Turin, Italy and employs over 137,801 people [21]. Its 2008 revenues were almost € 59 billion, 3.4% of which were invested in R&D. Fiat's Research Center (CRF) can be appropriately defined as the "innovation engine" of the Fiat Group, as it is responsible for the applied research and technology development activities of all its controlled companies [22]. The group Fiat has a diversified business portfolio, which shields it against demand fluctuations in certain product categories and also enables it to benefit from opportunities available in various divisions.

4.2. Statement of the problem and methodology

The aim of the project is to increase the flexibility and productivity in an ETU (Elementary Technology Unit) of Mechanical Subgroups in a part of the FGA's assembling process in the Cassino Plant through the conventional Plan-Do-Check-Act approach using the WCM methodology:

- **PLAN** - Costs Analysis and Losses Analysis starting from Cost Deployment (CD) for the manufacturing process using the items and tools of Workplace Organization (WO) and for the handling process the Logistic and Customer Services (LOG) applications.

- **DO** - Analysis of the non-value-added Activities; analysis of re-balancing line and analysis of re-balancing of work activities in accordance with the analysis of the logistics flows using the material matrix and the flows matrix. Study and realization of prototypes to improve workstation ergonomics and to ensure minimum material handling; Application of countermeasures found in the production process and logistics (handling).

- CHECK – Analysis of results in order to verify productivity improvement, ergonomic improvement (WO) and the optimization of the internal handling (in the plant) and external logistics flows (LOG). Check of the losses reduction according to Cost Deployment (CD).

- ACT - Extension of the methodology and other cases.

Here below is a description of the Statement of the Problem and methodology.

4.2.1. PLAN: Costs analysis and losses analysis (CD) for the manufacturing process (WO) and for the handling process (LOG)

In this first part (PLAN) were analyzed the losses in the assembly process area so as to organize the activities to reduce the losses identified in the second part of the analysis (DO). Object of the study was the Mechanical Subgroups ETU - Elementary Technology Unit (in a part of the Cassino Plant Assembly Shop). The aim of this analysis was to identify a program allowing to generate savings policies based on Cost Deployment:

- Identify relationships between cost factors, processes generating costs and various types of waste and losses;

- Find relationships between waste and losses and their reductions.

In fact, in general a production system is characterized by several waste and losses (MUDA), such as:

- Non-value-added activities;

- Low balancing levels;

- Handling losses;

- Delay in material procurement;

- Defects;

- Troubleshooting Machines;

- Setup;

- Breakdown.

It is important to give a measure of all the losses identified in process examination. The data collection is therefore the *"key element"* for the development of activities of Cost Deployment. Here below in Figure 7 an example of losses identified from CD from the Assembly Shop is shown and in Figure 8 is shown an example of CD data collection regarding NVAA (Non-Value-Added Activities) for WO (for this case study we excluded check and rework losses) in the Mechanical Subgroups area. Finally Figure 9 shows Analysis of losses Cost Deployment.

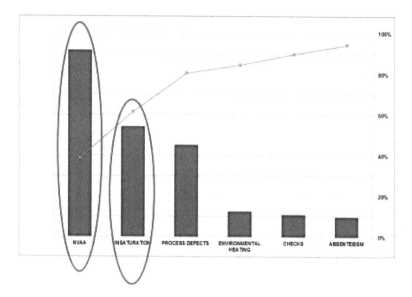

Figure 7. Analysis of losses Cost Deployment – Stratification of NVAA losses for Mechanical Subgroups ETU - Elementary Technology Unit (figure highlights the most critical workstation)

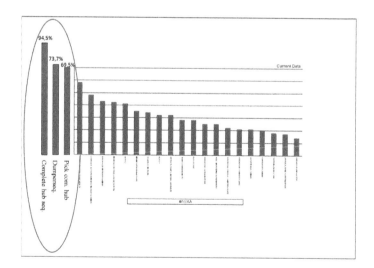

Figure 8. Analysis of losses Cost Deployment – Pareto Analysis NVAA Mechanical Subgroups ETU - Elementary Technology Unit

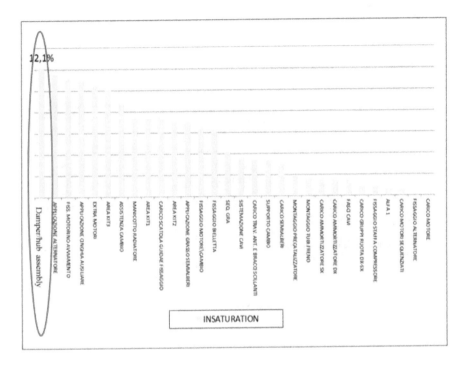

Figure 9. Analysis of losses Cost Deployment – Pareto Analysis Line Balancing Losses or Insaturation on Mechanical Subgroups ETU - Elementary Technology Unit

4.2.2. DO - Analysis of non-value-added activities, of the re-balancing line and analysis of re-balancing of work activities

According to figure 9 and figure 10 were analyzed the losses regarding NVAA and Insaturation. In fact were analyzed all 4 critical workstations (because they have the worst losses) and were identified 41 types of non-value-added activities (walking, waiting, turning, picking....) in the various sub-phases of the production process. In Table 4 is shown some examples of non-value-added activities analyzed (MUDA Analysis).

Some examples of standard tools used to analyze NVAA reduction (MUDA Analysis) for the 4 workstations are shown here below in Figures 10, 11 and 12) job stratification (VAA - Value Added Activities; NVAA – Non-Value-Added Activities; LBL - Low Balancing Level; EAWS - European Assembly Work Sheet – Ergonomy); 2) Spaghetti Chart and 3) Kaizen Standard.

N°	Losses identified	Solution	Non-value-added activities identified	
1	Pick picking list for sequencing	Complete hub sequencing	To pick	
2	Pick the box for sequencing	Complete hub sequencing	To pick	
3	Select sheets for the different model	Unification of the sheets from 3 to 1	To select	
4	Pick sheets for the different process	Unification of the sheets from 3 to 1	To pick	
5	Pick identification sheet	Unification of the sheets from 3 to 1	To pick	
6	Go to the printer to pick up sticker	Print sticker	To walk	
7	Pick identification hub label	Digital label with barcode	To pick	
8	Throw liner nameplate into the waste container	Print labels directly onto sheet unified	To trow	
9	Pick equipment for reading labels coupling	Automatic reading	To pick	
10	Combination of manual pallet	Automatic combination	To check	

N°	Losses identified	Solution	Non-value-added activities identified	
11	Use of a single box	Enabling a second workstation	To walk	
12	Pick hub	Pick subgroup (hub+ damper	To pick	
13	Use of electrical equipment through keyboard	New air equipment without keyboard	To arrange	
14	Use of air equipment through keyboard	New air equipment without keyboard	To wait	
15	Transport empty box hub sequencing to put the full box	Complete hub sequencing	To transport	
16	Walk to the line side to pick damper	Complete hub sequencing	To walk	
17	Remove the small parts to pair with damper	Complete hub sequencing	To pick	
18	Transport empty box damper sequencing to put the full box	Complete damper sequencing	To transport	
19	Pick the hub and put on the line	Pick subgroup (hub+ damper)	To pick	
20	Select the work program for the next workstation	Select the work program	To select	

N°	Losses identified	Solution	Non-value-added activities identified	
21	Press the feed button for the damper	Use a single workstation after the sequencing of the subgroup in order to press a button once	To push	
22	Wait for the translational motion of the pallet	Use a single workstation after the sequencing of the subgroup and match processing activities during the translation of the pallet	To wait	

Table 4. MUDA Analysis - NVAA

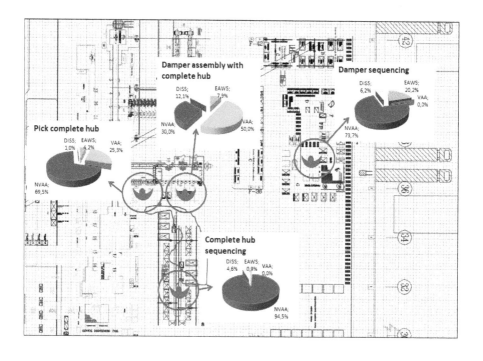

Figure 10. Details of the 4 workstations

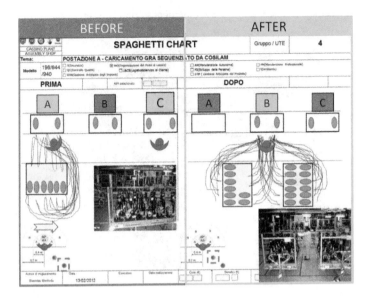

Figure 11. Spaghetti Chart Example

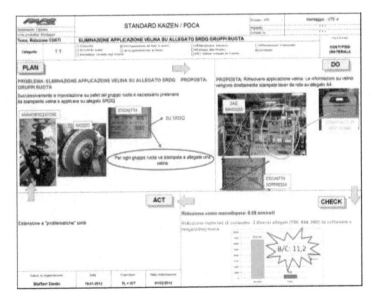

Figure 12. Standard Kaizen analysis Example

Figure 13 shows the initial scenario analyzed to identify problems and weaknesses.

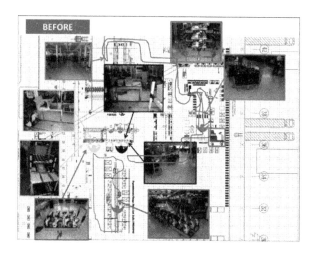

Figure 13. Details of the 4 workstations

At this point was assumed the new flow of the complete damper (corner) = damper + complete hub sequencing according to the material matrix considering losses relating to handling (material matrix classification – see figure 14). The material matrix classifies the commodities (number of drawings) in three main groups: A (bulky, multi-variations, expensive), B (normal) and C (small parts) and subgroups (a mixture of group A: bulky and multi-variations or bulky and expensive etc.). For each of these groups was filled out the flow matrix that defines the correct flow associated: JIS (and different levels), JIT (and different levels) and indirect (and different levels). After identifying the correct flow, in the JIS case, was built a prototype of the box (bin) to feed the line that would ensure the right number of parts to optimize logistic handling. However, the new box (bin) for this new mechanical subgroup must feed the line in a comfortable and ergonomic manner for the worker in the workstation, for this reason was simulated the solution before the realization of the box (bin) (see figure 15).

At the end of the Muda analysis (NVAA analysis) were applied all the solutions found to have a lean process (the internal target is to achieve 25% of average NVAA losses) and was reorganized the line through a new line balancing level (rebalancing) to achieve 5% of the average line balancing losses (internal target). Another important aspect was the logistics flows analysis (see figure 16) considering *advanced warehouses* (Figure 17). The simulation scenario was defined using trucks from the Cassino plant warehouses that also feed other commodities to achieve high levels of saturation to minimize handling losses.

At the end of the handling analysis (flow, stock level…) thanks to this new "lean" organization of material matrix was used the correct line feed from the Just In Sequence warehouse.

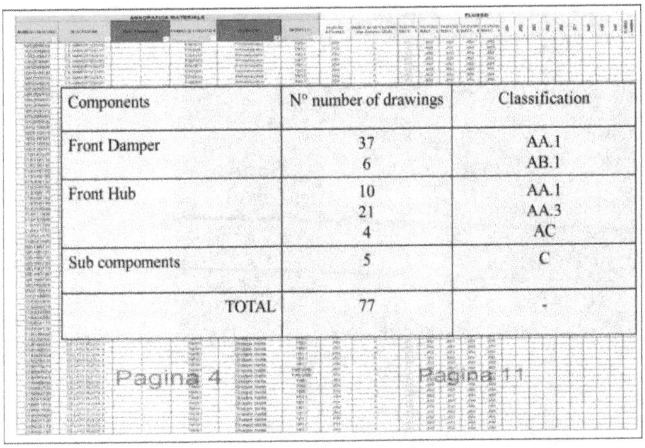

Components	N° number of drawings	Classification
Front Damper	37	AA.1
	6	AB.1
Front Hub	10	AA.1
	21	AA.3
	4	AC
Sub compoments	5	C
TOTAL	77	-

Figure 14. Material matrix example

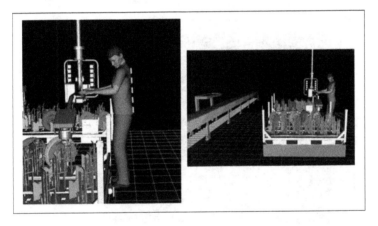

Figure 15. Simulation of an ergonomic workstation

It was reduced the internal warehouse (stock level), the space used for sequencing (square metres), the indirect manpower used to feed the sequencing area and we obtained zero fork-lifts on the shopfloor because we used the ro-ro (roll in - roll out) system. Figure 18 shows the final scenario in which we have 1 operator instead of 4 operators.

4.2.3. Check – Analysis of results to verify productivity and ergonomic improvement and optimization of logistics flows

In detail the main results and savings can be summarized as follows:

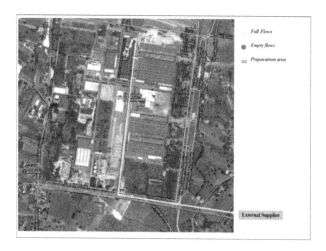

Figure 16. Initial logistic flows

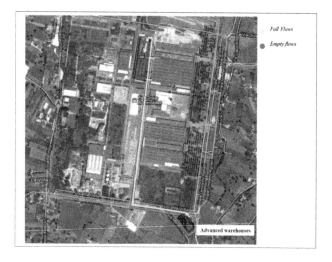

Figure 17. Logistic flows considering advanced warehouses

- Productivity improvement +75% (Figure 19) direct labour;

- Ergonomics improvement +85% (Figure 20) according to the rest factor;

- Optimization of logistic flows (Figure 21) according to the flow matrix.

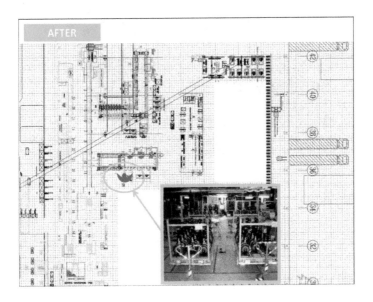

Figure 18. Details of the final workstation

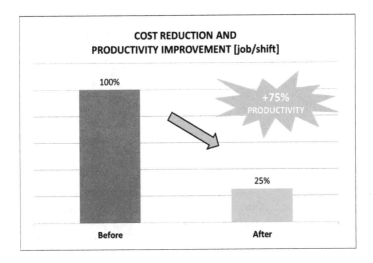

Figure 19. Productivity optimization

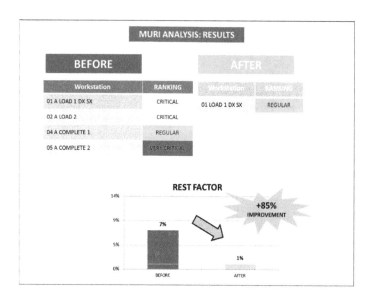

Figure 20. Ergonomics improvement

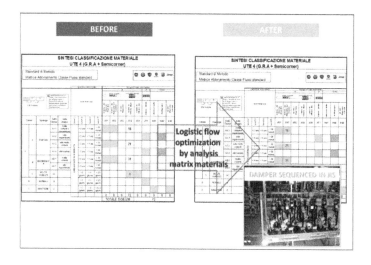

Figure 21. Optimization of logistic flows

4.2.4. Act - Extension of the methodology and other cases

Future developments include the extension of the methodology to the entire plant. Here below in Table 5 we can see the activities and status adopted. to achieve the results shown in the "check". We used traditional tools and methodology for the analysis and new tools to simulate the sceneries on the line and for the logistic problems we involved other resources outside the plant (ELASIS and CRF - FIAT Research Center, Fiat Central Department and Public Universities).

ACTIVITIES	TOOL	STATUS
NVAA Reduction	NVAA Std analysis	+
	NVAA Database	+
LBL Reduction	Balance line	+
	Check Saturation (Flexim Software/Plant simulation)	+
Ergonomics Improvement	Jack Software	+
	Excel Human Model	+
Optimization of logistics flow	Value stream map	+
	Check Saturation Flexim/Plant simulation	+

Table 5. Activities and status

5. Conclusions

A key industrial policy conclusion is that intelligently designed selective policies can be effective in developing production systems. Intelligent industrial policies need to be shaped to respond to contingent factors which are specific to a sector, period and country. Fundamentally, it is not a question of whether these selective policies work, but under what circumstances they work.

From this point of view, World Class Manufacturing is a "key" concept. This is the reason why the concept constituting "World Class Manufacturing" has received considerable attention in academic literature, even though it has been developed principally in relation to the needs of larger scale manufacturing organisations. Regards our case study we can conclude that WCM allows to reduce losses and optimize logistics flows. Thus, the main results can be summarized as follows:

1. greater efficiency because the inner product is cheaper because it is possible to use external warehouses or suppliers - outsourcing - specialized and more cost-effective for the company;

2. greater flexibility because it is possible to work more models (in Cassino with these logical sequencing and kitting there are 4 different model brands on the same assembly line: *Alfa Romeo Giulietta, Chrysler, Lancia Delta and Fiat Bravo*;

3. no space constraint (in this example we get only 1 container already sequenced line side)

Definitely the new process and the internal flows are very lean and efficient. In this case study it was implemented a servo system using Low Cost Automation. This system ensures only one picking point in order to have only one container at the side of the production line.

Acknowledgements

We would like to express our gratitude to Fiat Group Automobiles S.p.A. - Cassino Plant, to the Plant Manager and his staff who gave us the possibility to carry out the necessary research and use their data.

Author details

Fabio De Felice[1], Antonella Petrillo[1*] and Stanislao Monfreda[2]

*Address all correspondence to: a.petrillo@unicas.it

1 University of Cassino, Department of Civil and Mechanical Engineering, Cassino, Italy

2 Fiat Group Automobiles EMEA WCM Cassino Plant Coordinator, Cassino, Italy

References

[1] De Felice F., Petrillo A, Silvestri, A. Multi-criteria risk analysis to improve safety in manufacturing systems. International Journal of Production Research. 2012; Vol. 50, No. 17, pp. 4806-4822.

[2] De Felice F., Petrillo A. Methodological Approach for Performing Human Reliability and Error Analysis in Railway Transportation System. International Journal of Engineering and Technology 2011; Vol.3 (5), 341-353.

[3] De Felice F., Petrillo A. Hierarchical model to optimize performance in logistics policies: multi attribute analysis. The 8th International Strategic Management Conference. June 21-23, 2012 Barcelona –Spain. Elsevier Procedia Social and Behavioral Sciences.

[4] De Felice F., Petrillo A. Productivity analysis through simulation technique to optimize an automated assembly line. Proceedings of the IASTED International Conference, June 25 - 27, 2012 Napoli, Italy. Applied Simulation and Modelling (ASM 2012) DOI: 10.2316/ P.2012.776-048 – pp 35-42.

[5] Haynes A. Effect of world class manufacturing on shop floor workers, Journal European Industrial Training 1999; 23(6) 300–309.

[6] Womack J. P., Jones D. T., Roos D. The Machine that Changed the World (Rawson Associates, New York, 1990).

[7] Oliver N., Delbridge R., Jones D., and Lowe J. World class manufacturing: Further evidence in the lean production debate, British Journal of Management 5(Special issue) (1994) S53–S63.

[8] Schoenberger R.J. World class manufacturing: the lessons of simplicity applied, New York: Free Press, p. 205, 1986.

[9] Yamashina H. Japanese manufacturing strategy and the role of total productive maintenance. Journal of Quality in Maintenance Engineering Volume 1, Issue 1, 1995, Pages 27-38.

[10] Ghalayini A.M., and Noble J.S. The changing basis of performance measurement, Int. J. Operations & Production Management 16(8) (1996) 63–80.

[11] Kodali R.B., Sangwan K.S., Sunnapwar V.K. Performance value analysis for the justification of world-class manufacturing systems, J. Advanced Manufacturing Systems 3(1) (2004) 85–102.

[12] Wee Y.S., and Quazi H.A. Development and validation of critical factors of environmental management, Industrial Management & Data Systems 105(1) (2005) 96–114.

[13] Digalwar A.K., Metri, B.A. Performance measurement framework for world class manufacturing, International Journal Applied Management & Technology 3(2) (2005) 83–102.

[14] Utzig L. J. CMS performance measurement, in Cost Management for Today's Advanced Manufacturing: The CAM-I Conceptual Design, eds. C. Berliner and J. A. Brimson (Harvard Business School Press, Boston, 1988).

[15] Hayes R.H., Abernathy W. J. Managing our way to economic decline, Harvard Business Review 58(1) (1980) 67–77.

[16] Schmenner R. W. International factory productivity gains, J. Operations Management 10(2) (1991) 229–54.

[17] Kennerley M., Neely A., Adams C. Survival of the fittest measuring performance in a changing business environment, Measuring Business Excellence 7(4) (2003) 37–43.

[18] Japan Institute of Plant Maintenance (JIPM) (ed.): A Report on Systemizing Indicators of Total Productive Maintenance (TPM) (in Japanese) (JIPM, Tokyo, 2007).

[19] Shirose K. (ed.): TPM New Implementation Program in Fabrication and Assembly Industries Productivity Press, Portland, Oregon 1996.

[20] Murata K., Katayam H. 2009. An evaluation of factory performance utilized KPI/KAI with data envelopment analysis Journal of the Operations Research Society of Japan 2009, Vol. 52, No. 2, 204-220.

[21] Datamonitor. Fiat S.p.A. Company Profile. 12 July 2011.

[22] Di Minin A., Frattini F., Piccaluga A. Fiat: Open Innovation in a downturn (1993-2003). University of California, Berkeley vol., 52 (3). Spring 2010.

Chapter 2

Using Overall Equipment Effectiveness for Manufacturing System Design

Vittorio Cesarotti, Alessio Giuiusa and Vito Introna

Additional information is available at the end of the chapter

1. Introduction

Different metrics for measuring and analyzing the productivity of manufacturing systems have been studied for several decades. The traditional metrics for measuring productivity were *throughput* and *utilization rate*, which only measure part of the performance of manufacturing equipment. But, they were not very helpful for *"identifying the problems and underlying improvements needed to increase productivity"* [1].

During the last years, several societal elements have raised the interest in analyze the phenomena underlying the identification of productive performance parameters as: capacity, production throughput, utilization, saturation, availability, quality, etc.

This rising interest has highlighted the need for more rigorously defined and acknowledged productivity metrics that allow to take into account a set of synthetic but important factors (availability, performance and quality) [1]. Most relevant causes identified in literature are:

- The growing attention devoted by the management to cost reduction approaches [2] [3];

- The interest connected to successful eastern productions approaches, like Total *Productive Maintenance* [4], *World Class Manufacturing* [5] or *Lean production* [6];

- The importance to go beyond the limits of traditional business management control system [7];

For this reasons, a variety of new performance concepts have been developed. The total productive maintenance (TPM) concept, launched by Seiichi Nakajima [4] in the 1980s, has provided probably the most acknowledged and widespread quantitative metric for the measure of the productivity of any production equipment in a factory: the *Overall Equipment Effectiveness* (OEE). OEE is an appropriate measure for manufacturing organizations and it has being used broadly in manufacturing industry, typically to monitor and control the performance (time losses) of an equipment/work station within a production system [8]. The OEE allows to quantify and to assign all the time losses, that affect an equipment whilst the production, to three standard categories. Being standard and widely acknowledged, OEE has constituted a powerful tool for production systems performance benchmarking and characterization, as also the starting point for several analysis techniques, continuous improvement and research [9] [10]. Despite this widespread and relevance, the use of OEE presents limitations. As a matter of fact, OEE focus is on the single equipment, yet the performance of a single equipment in a production system is generally influenced by the performance of other systems to which it is interconnected. The time losses propagation from a station to another may widely affect the performance of a single equipment. Since OEE measures the performance of the equipment within the specific system, a low value of OEE for a given equipment can depend either on little performance of the equipment itself and/or time losses propagation due to other interconnected equipments of the system.

This issue has been widely investigated in literature through the introduction of a new metric: the Overall Equipment Effectiveness (OTE), that considers the whole production system as a whole. OTE embraces the performance losses of a production system both due to the equipments and their interactions.

Process Designers need usually to identify the number of each equipments necessary to realize each activity of the production process, considering the interaction and consequent time losses a priori. Hence, for a proper design of the system, we believe that the OEE provides designer with better information on each equipment than OTE. In this chapter we will show how OEE can be used to carry out a correct equipments sizing and an effective production system design, taking into account both equipment time losses and their propagation throughout the whole production system.

In the first paragraph we will show the approach that a process designer should face when designing a new production system starting from scratch.

In the second paragraph we will investigate the typical time-losses that affect a production system, although are independent from the production system itself.

In the third part we will define all the internal time losses that need to be considered when assessing the OEE, along with the description of a set of critical factors related to OEE assessment, such as buffer-sizing and choice of the plant layout.

In the fourth paragraph we will show and quantify how time losses of a single equipment affects the whole system and vice-versa.

Finally, we will show through the simulation some real cases in which a process design have been fully completed, considering both equipment and time losses propagation.

2. Manufacturing system design: Establish the number of production machines

Each process designer, when starting the design of a new production system, must ensure that the number of equipments necessary to carry out a given process activity (e.g. metal milling) is sufficient to realize the required volume. Still, the designer must generally ensure that the minimum number of equipment is bought due to elevated investment costs. Clearly, the performance inefficiencies and their propagation became critical, when the purchase of an extra (set of) equipment(s) is required to offset time losses propagation. From a price strategy perspective, the process designer is generally requested to assure the number of requested equipments is effectively the minimum possible for the requested volume. Any not necessary over-sizing results in an extra investment cost for the company, compromising the economical performance.

Typically, the general equation to assess the number of equipments needed to process a demand of products (D) within a total calendar time C_t (usually one year) can be written as follow (1):

$$n_i = int\left[\frac{D^*ct_i}{C_t{}^*\vartheta^*\eta_i}\right] + 1 \tag{1}$$

Where:

• D is the number of products that must be produced;

• ct_i is theoretical cycle time for the equipment i to process a piece of product;

• C_t is the number of hours (or minutes) in one year.

• ϑ is a coefficient that includes all the external time losses that affect a production system, precluding production.

• η_i is the efficiency of the equipment i within the system.

It is therefore possible to define L_t, Loading time, as the percentage of total calendar time C_t that is actually scheduled for operation (2):

$$L_t = C_t{}^*\vartheta \tag{2}$$

The equation (1) shows that the process designer must consider in his/her analysis three parameters unknown a priori, which influence dramatically the production system sizing and play a key role in the design of the system in order to realize the desired throughput. These parameters affect the total time available for production and the real time each equipment request to realize a piece [9], and are respectively:

• External time losses, which are considered in the analysis with ϑ ;

- The theoretical time cycle, which depends upon the selected equipment(s);

- The efficiency of the equipment which depends upon the selected equipments and their interactions, in accordance to the specific design.

This list highlights the complexity implicitly involved in a process design. Several forecasts and assumptions may be required. In this sense, it is a good practice to ensure that the ratio in equation (3) is always respected for each equipment:

$$\frac{\left(\frac{D^*ct_i}{L_t{\cdot}\eta_i}\right)}{n_i} < 1 \qquad (3)$$

As a good practice, to ensure (3) being properly lower than 1 allows to embrace, among others, the variability and uncertainty implicitly embedded within the demand forecast.

In the next paragraph we will analyze the External time losses that must be considered during the design.

3. External time losses

3.1. Background

For the design of a production system several time-losses, of different nature, need to be considered. Literature is plenty of classifications in this sense, although they can diverge one each others in parameters, number, categorization, level of detail, etc. [11] [12]. Usually each classification is tailored on a set of sensible drivers, such as data availability, expected results, etc. [13].

One relevant classification of both external and internal time losses is provided by Grando et al. [14]. Starting from this classification and focusing on external time losses only, we will briefly introduce a description of common time-losses in Operations Management, highlighting which are most relevant and which are negligible under certain hypothesis for the design of a production system (Table 1).

The categories LT1 and LT2 don't affect the performance of a single equipment, nor influence the propagation of time-losses throughout the production system.

Still, it is important to notice that some causes, even though labeled as external, are complex to asses during the design. Despite these causes are external, and well known by operations manager, due to the implicit complexity in assessing them, these are detected only when the production system is working via the OEE, with consequence on OEE values. For example, the lack of material feeding a production line does not depend by the OEE of the specific station/equipment. Nevertheless when lack of material occurs a station cannot produce with consequences on equipment efficiency, detected by the OEE. (4).

Symbol	Name	Description	Synonyms
Lt1	Idle times resulting from law regulations or corporate decisions	Summer vacations, holidays, shifts, special events (earthquakes, flood);	System External Causes
Lt2	Unplanned time	Lack of demand; Lack of material in stocks;	System External Causes
		Lack of orders in the production plan;	
		Lack of energy; Lack of manpower (strikes, absenteeism);	
		Technical tests and manufacturing of nonmarketable products; Training of workers;	
Lt3	Stand by time	Micro-absenteism, shift changes; physiological increases; man machine interaction;	Machine External Causes; System External Causes
		Lack of raw material stocks for single machines; Unsuitable physical and chemical properties of the available material;	
		Lack of service vehicle; Failure to other machines;	

Table 1. Adapted from Grando et al. 2005

3.2. Considerations

The external time losses assessment may vary in accordance to theirs categories, historical available data and other exogenous factors. Some stops are established for through internal policies (e.g. number of shift, production system closure for holidays, etc.). Other macro-stops are assessed (e.g. Opening time to satisfy forecasted demand), whereas others are considered as a forfeit in accordance to the Operations Manager Experience. It is not possible to provide a general magnitude order because, the extent of time losses depend from a variety of characteristic factor connected mainly to the specific process and the specific firm. Among the most common ways to assess this time losses we found: Historical data, Benchmarking with similar production system, Operations Manager Experience, Corporate Policies.

The Calendar time C_t is reduced after the external time losses. The percentage of C_t in which the production system does not produce is expressed by (1- ϑ) , affecting consequently the L_t (2).

These parameters should be considered carefully by system designers in assessing the loading time (2). Although these parameters do not propagate throughout the line their consideration is fundamental to ensure the identification of a proper number of equipments.

3.2.1. Idle times

There is a set of idle times that result from law regulations or corporate decisions. These stops are generally known a-priori, since they are regulated by local law and usually contribute to the production plant localization-decision process. Only causes external to the production system are responsible for their presence.

3.2.2. Unplanned times

The unplanned time are generally generated by system external causes connected with machineries, production planning and production risks.

A whole production system (e.g. production line) or some equipment may be temporarily used for non marketable product (e.g. prototype), or they may are not supposed to produce, due to test (e.g. for law requirements), installation of new equipments and the related activities (e.g. training of workers).

Similarly, a production system may face idle time because of lack of demand, absence of a production schedule (ineffectiveness of marketing function or production planning activities) or lack of material in stock due to ineffectiveness in managing the orders. Clearly, the presence of a production schedule in a production system is independent by the Operations manager and by the production system design as well. Yet, the lack of stock material, although independent from the production system design is one of the critical responsibility of any OM (inventory management).

Among this set of time losses we find also other external factors that affect the system availability, which are usually managed by companies as a risk. In this sense occurrence of phenomenon like the lack of energy or the presence of strikes are risks that companies well know and that usually manage according to one of the four risk management strategy (avoidance, transfer, mitigation acceptance) depending on their impact and probability.

3.2.3. Stand by time

Finally, the stand-by time losses are a set of losses due to system internal causes, but still equipment external causes. This time losses may affect widely the OTE of the production line and depend on: work organization losses, raw material and material handling.

Micro-absenteeism and shift changes may affect the performances of all the system that are based on man-machine interaction, such as the production equipments or the transportation systems as well. Lack of performance may propagate throughout the whole system as other equipment ineffectiveness. Even so, Operations manager can't avoid these losses by designing a better production line. Effective strategies in this sense are connected with social science that aim to achieve the employee engagement in the workplace [15].

Nonetheless Operations Manager can avoid the physiological increases by choosing ergonomic workstations.

The production system can present other time-losses because of the raw material, both in term of lack and quality:

- Lack of raw material causes the interruption of the throughput. Since we have already considered the ineffective management of the orders in "Unplanned Time", the other related causes of time-losses depend on demand fluctuation or in ineffectiveness of the suppliers as well. In both cases the presence of safety stock allows operations manager to reduce or eliminate theirs effects.

- Low raw material standard quality (e.g. physical and chemical properties), may affect dramatically the performance of the system. Production resource (time, equipment, etc) are used to elaborate a throughput without value (or with a lower value) because of little raw material quality. Also in this case, this time losses do not affect the design of a production system, under the hypothesis that Operations Manager ensures the raw material quality is respected (e.g. incoming goods inspection). The missed detection of low quality raw materials can lead the Operations Manager to attribute the cause of defectiveness to the equipment (or set of equipment) where the defect is detected.

Considering the Vehicle based internal transport, a broader set of considerations is requested. Given two consecutive stations i-j, the vehicles make available the output of station i to station j (figure 1).

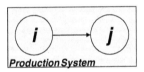

Figure 1. Vehicle based internal transport: transport the output of station i to the station j

In this sense any vehicle can be considered as an equipment that is carrying out the transformation on a piece, moving the piece itself from station i to station j (Figure 2).

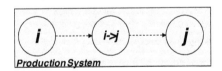

Figure 2. Service vehicles that connect i-j can be represented as a station itself amid i-j

The activity to transport the output from station i to station j is a transformation (position) itself. Like the equipments, also the service vehicles affect and are affected by the OTE. In this sense successive considerations on equipments losses categorization, OEE, and their propagations throughout the system, OTE, can be extended to service vehicles. Hence, the design of service vehicles would be carried out according to the same guidelines we provide in successive section of this chapter.

4. The formulation of OEE

In this paragraph we will provide process designer with a set of topics that need to be addressed when considering the OEE during the design of a new production system. A proper assessment a-priori of the OEE, and the consequent design and sizing of the system demand process designer to consider a variety of complex factors, all related with OEE. It is important to notice that OEE measures not only the internal losses of efficiency, but is also detects time losses due to external time losses (par.2.1, par.2.2). Hence, in this paragraph we will firstly define analytically the OEE. Secondly we will investigate, through the analysis of relevant literature, the relation between the OEE of single equipment and the OEE of the production system as a set of interconnected equipments. Then we will describe how different time losses categories, of an equipment, affect both the OEE of the equipment and the OEE of the Whole system. Finally we will debate how OEE need to be considered with different perspective in accordance to factors as ways to realize the production and plant layout.

4.1. Mathematical formulation

OEE is formulated as a function of a number of mutually exclusive components, such as *availability efficiency, performance efficiency,* and *quality efficiency* in order to quantify various types of productivity losses.

OEE is a value variable from 0 to 100%. An high value of OEE indicates that machine is operating close to its maximum efficiency. Although the OEE does not diagnose a specific reason why a machine is not running as efficiently as possible, it does give some insight into the reason [16]. It is therefore possible to analyze these areas to determine where the lack of efficiency is occurring: breakdown, set-up and adjustment, idling and minor storage, reduced speed, and quality defect and rework [1] [4].

In literature exist a meaningful set of time losses classification related to the three reported efficiencies (availability, performance and quality). Grando et al. [14] for example provided a meaningful and comprehensive classification of the time-losses that affect a single equipment, considering its interaction in the interaction system. Waters et al. [9] and Chase et al. [17] showed a variety of acknowledged possible efficiency losses schemes, while Nakajima [4] defined the most acknowledged classification of the "6 big losses".

In accordance with Nakajima notations, the conventional formula for OEE can be written as follow [1]:

$$OEE = A_{eff} \ Pe_{eff} \ Q_{eff} \tag{4}$$

$$A_{eff} = \frac{T_u}{T_t} \tag{5}$$

$$Pe_{eff} = \frac{T_p}{T_u} * \frac{R_{avg}^{(a)}}{R_{avg}^{(th)}} \tag{6}$$

$$Q_{eff} = \frac{P_g}{P_a} \qquad (7)$$

Table 2 summarizes briefly each factor.

Factor	Description
A_{eff}	Availability efficiency. It considers failure and maintenance downtime and time devoted to indirect production task (e.g. set up, changeovers).
Pe_{eff}	Performance efficiency. It consider minor stoppages and time losses caused by speed reduction
Q_{eff}	Quality efficiency. It consider loss of production caused by scraps and rework.
T_u	Equipment uptime during the T_t. It is lower that T_t because of failure, maintenance and set up.
T_t	Total time of observation.
T_p	Equipment production time. It is lower than T_t because of minor stoppages, resets, adjustments following changeovers.
$R_{avg}^{(a)}$	Average actual processing rate for equipment in production for actual product output. It is lower than theoretical ($R_{avg}^{(th)}$) because of speed/production rate slowdowns.
$R_{avg}^{(th)}$	Average theoretical processing rate for actual product output.
P_g	Good product output from equipment during T_t.
P_a	Actual product units processed by equipment during T_t. We assume that for each product rework the same cycle time is requested.

Table 2. OEE factors description

The OEE analysis, if based on single equipment data, is not sufficient, since *no machine is isolated in a factory, but operates in a linked and complex environment* [18]. A set of inter-dependent relations between two or more equipments of a production system generally exists, which leads to the propagation of availability, performance and quality losses throughout the system.

Mutual influence between two consecutive stations occurs even if both stations are working ideally. In fact if two consecutive stations (e.g. station A and station B) present different cycle times, the faster station (eg. Station A = 100 pcs/hour) need to reduce/stop its production rate in accordance with the other station production rate (e.g. Station B = 80 pcs/hour).

Station A	Station B
100 pcs/hour	80 pcs/hour

In this case, the detected OEE of station A would be 80%, even if any efficiency loss occurs. This losses propagation is due to the unbalanced cycle time.

Therefore, when considering the OEE of equipment in a given manufacturing system, the measured OEE is always the performance of the equipment within the specific system. This leads to practical consequence for the design of the system itself.

A comprehensive analysis of the production system performance can be reached by extending the concept of OEE, as the performance of individual equipment, up to factory level [18]. In this sense OEE metric is well accepted as an effective measure of manufacturing performance not only for single machine but also for the whole production system [19] and it is known as *Overall Throughput Effectiveness* OTE [1] [20].

We refer to OTE as the OEE of the whole production system.

Therefore we can talk of:

- Equipment OEE, as the OEE of the single equipment, which measures the performance of the equipment in the given production system.

- System OEE (or OTE), which is the performance of the whole system and can be defined as the performance of the bottleneck equipment in the given production system.

4.2. An analytical formulation to study equipment and system OEE

$$\text{System OEE} = \frac{\text{Number of good parts produced by system in total time}}{\text{Theoretical number of parts produced by system in total time}} \qquad (8)$$

The System OEE measures the systemic performance of a manufacturing system (productive line, floor, factory) which combines activities, relationships between different machines and processes, integrating information, decisions and actions across many independents systems and subsystem [1]. For its optimization it is necessary to improve coordinately many interdependent activities. This will also increase the focus on the plant-wide picture.

Figure 3 clarify which is the difference between Equipment OEE and System OEE, showing how the performance of each equipment affects and is affected by the performances of the other connected equipments. These time losses propagation result on a Overall System OEE. Considering the figure 3 we can indeed argue that given a set of $i=1,..,n$ equipments, OEE_i of the i th equipment depends on the process in which it has been introduced, due to the availability, performance and quality losses propagation.

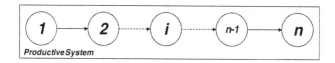

Figure 3. A production system composed of n stations

According to the model proposed by Huang et al in [1], the System OEE (OTE) for a series of n connected subsystems, is formulated in function of theoretical production rate $R_{avg(F)}^{(th)}$ relating to the slowest machine (the bottleneck), theoretical production rate $R^{(th)}_{avg(N)}$ and OEE_n of nth station as shown in (9):

$$OTE = \frac{OEE_n \times R^{th}_{avg(n)}}{R^{th}_{avg(F)}} \tag{9}$$

The OEE_n computed in (9) is the OEE of nth station introduced in the production system (the OEE_n when n is in the system and it is influenced by the performance of other n-1 equipments).

According to (9) the only measure of OEE_n is a measure of the performance of the whole system (OTE). This is true because performance data on n are gathered when the station n is already working in the system with the other n-1 station and, therefore, its performance is affected from the performance of the other n-1 prior stations. This means that the model proposed by Huang, could be used *only when the system exists and it is running,* so OEE_n could be directly measured on field.

But during system design, when only technical data of single equipment are known, the same formulation in (9) can't be used, since without information on the system OEE_n in unknown a-priori. Hence, in this case the (9) couldn't provide a correct value of OTE.

4.3. How equipment time-losses influence the system performance and vice-versa

The OEE of each equipment, *as isolated machine* (independent by other station) is affected only by (5),(6) and (7) theoretical intrinsic value. But once the equipment is part of a system its performance depends also upon the interaction with other n-1 equipments and thus on their performance. It is now more evident why, for a correct estimate and/or analysis of equipment OEE and system OEE, it is necessary to take into account losses propagation. These differences between single subsystem and entire system need to be deeply analyzed to understand real causes of system efficiency looses. In particular their investigation is fundamental during the design process, because a correct evaluation of OEE and for the study of effective losses reduction actions (i.e. buffer capacity dimensioning, quality control station positioning); but also during the normal execution of the operations because it leads to correct evaluation of causes of efficiency losses and their real impact on the system.

The table 3 shows how efficiency losses of a single subsystem (e.g. an equipment/ machine), given by Nakajima [4] can spread to other subsystem (e.g. in series machines) and then to whole system.

In accordance to table 3 a relevant lack of coordination in deploying available factory resources (people, information, materials, and tools) by using OEE metric (based on single equipment) exists. Hence, a wider approach for a holistic production system design has to focus also *on the performance of the whole factory* [18], resulting by the interactions of its equipments.

	Single subsystem	Entire system
Availability	Breakdown losses	Downtimes losses of upstream unit could slackening production rate
	Set-up and adjustment	of downstream unit without fair buffer capacity
		Downtimes losses of downstream unit could slackening production
		rate of upstream unit without fair buffer capacity
Performance	Idling and minor stoppages	Minor stoppages and speed reduction could influencing production
	Reduced speed	rate of the downstream and upstream unit in absence of buffer
Quality	Quality defects and rework	Production scraps and rework are losses for entire process depends on
	Yield losses	where the scraps are identified, rejected or reworked in the process

Table 3. Example of propagation of losses in the system

This issue have been widely debated and acknowledged in literature [1] [18]. Several Authors [8] [21] have recognized and analyzed the need for a coherent, systematic methodology for design at the factory level.

Furthermore, the following activities, according to [18] [21] have to be considered as OTE is also started at the factory design level:

- Quality (better equipment reliability, higher yields, less rework, no misprocessing);

- Agility and responsiveness (more customization, fast response to unexpected changes, simpler integration);

- Technology changes;

- Speed (faster ramp up, shorter cycle times, faster delivery);

- Production cost (better asset utilization, higher throughput, less inventory, less setup, less idle time);

At present, there is not a common well defined and proven methodology for the analysis of System OEE [1] [19] *during the system design.* By the way the effect of efficiency losses propagation must be considered and deeply analyzed to understand and eliminate the causes before the production system is realized. In this sense the *simulation* is considered the most reliable method, to date, in designing, studying and analyzing the manufacturing systems and its dynamic performance [1] [19]. Discrete event simulation and advanced process control are the most representatives of such areas [22].

4.4. Layout impact on OEE

Finally, it is important to consider how the focus of the design may vary according the type of production system. In flow-shop production system the design mostly focuses on the OTE of the whole production line, whereas in job-shop production system the analysis may focus either on the OEE of a single equipment or in those of the specific shop floor, rather than those of the whole production system. This is due to the intrinsic factors that underlies a layout configuration choice.

Flow shop production systems are typical of high volume and low variety production. The equipment present all a similar cycle time [23] and is usually organized in a product layout where interoperation buffers are small or absent. Due to similarity among the equipments that compose the production system, the saturation level of the different equipments are likely to be similar one each other. The OEE are similar as well. In this sense the focus of the analysis will be on loss time propagation causes, with the aim to avoid their occurrence to rise the OTE of the system.

On the other hand, in *job shop production systems*, due to the specific nature of operations (multi-flows, different productive paths, need for process flexibility rather than efficiency) characterized by higher idle time and higher stand-by-time, lower values of performances index are pursued.

Different products categories usually require a different sequence of tasks within the same production system so the equipment is organized in a process layout. In this case rather than focusing on efficiency, the design focuses on production system flexibility and in the layout optimization in order to ensure that different production processes can take place effectively.

Generally different processes, to produce different products, imply that bottleneck may shift from a station to another due to different production processes and different processing time of each station in accordance to the specific processed product as well.

Due to the shift of bottleneck the presence of buffers between the stations usually allows different stations to work in an asynchronous manner, consecutively reducing/eliminating the propagation of low utilization rates.

Nevertheless, when the productive mix is known and stable over time, the study of plant layout can embrace bottleneck optimization for each product of the mix, since a lower flexibility is demanded.

The analysis of quality propagation amid two or more stations should not be a relevant issue in job shop, since defects are usually detected and managed within the specific station.

Still, in several manufacturing system, despite a flow shop production, the equipment is organized in a process layout due to physical attributes of equipment (e.g. manufacturing of electrical cables showed in § 4) or different operational condition (e.g. pharmaceutical sector). In this case usually buffers are present and their size can dramatically influence the OTE of the production system.

In an explicit attempt to avoid unmanageable models, we will now provide process designers and operations managers with useful hints and suggestion about the effect of inefficiencies propagation among a production line along with the development of a set of simulation scenarios (§ 3.5).

4.5. OEE and OTE factors for production system design

OEE is formulated as a function of a number of mutually exclusive components, such as availability efficiency, performance efficiency, and quality efficiency in order to quantify various types of productivity losses.

During the design of the production system the use of intrinsic performance index for the sizing of each equipment although wrong could seem the only rational approach for the design. By the way, this approach don't consider the interaction between the stations. Someone can argue that to make independent each station from the other stations through the buffer would simplify the design and increase the availability. Still, the interposition of a buffer between two or more station may not be possible for several reason. Most relevant are:

• logistic (space unavailability, huge size of the product, compact plant layout, etc.);

• economic (the creation of stock amid each couple of station increase the WIP and consequently interest on current assets);

• performance;

• product features (buffer increase cross times, critical for perishable products);

In our model we will show how a production system can be defined considering availability, performance and quality efficiency (5),(6), (7) of each station along with their interactions. The method embraces a set of hints and suggestions (best practices) that lead designers in handle interactions and losses propagation with the aim to rise the expected performance of the system. Furthermore, through the development of a simulation model of a real production system for the electrical cable production we provide students with a clear understanding of how time-losses propagate in a real manufacturing system.

The design process of a new production system should always include the simulation of the identified solution, since the simulation provides designer with a holistic understanding of the system. In this sense in this paragraph we provide a method where the design of a production system is an iterative process: the simulation output is the input of a successive design step, until the designed system meet the expected performance and performance are validated by simulation. Each loss will be firstly described referring to a single equipment, than its effect will be analyzed considering the whole system, also throughout the support of simulation tools.

4.5.1. Set up availability

Availability losses due to set up and changeover must be considered during the design of the plant. In accordance with the production mix, the number of set-up generally results as a trade-off between the set up costs (due to loss of availability + substituted tools, etc.) and the warehouse cost.

During the design phase some relevant consideration connected with set-up time losses should be considered. A production line is composed of n stations. The same line can usually produce more than one product type. Depending on the difference between different product types a changeover in one or more stations of the line can be required. Usually, the more negligible the differences between the products, the lower the number of equipments subjected to set up (e.g. it is sufficient the set up only of the label machine to change the labels of a product depending on the destination country). In a given line of n equipments, if a set up is requested in station i, loss availability can interest only the single equipment I or the whole production line, depending on the buffer presence, their location and dimension:

- If buffers are not present, the set up of station i implies the stop of the whole line (figure 4). This is a typical configuration of flow shop process realized by one or more production line as food, beverages, pharmaceutical packaging,....

- If buffers are present (before and beyond the station i) and their size is sufficient to decouple the station i by the other i-1 and i+1 station during the whole set up, the line continues to work regularly (figure 5).

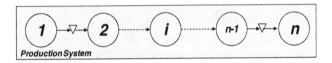

Figure 4. Barely decoupled/Coupled Production System (buffer unimportant or null)

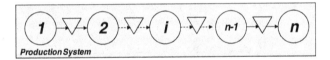

Figure 5. Decoupled Production System

Hence, the buffer design plays a key role in the phenomena of losses propagation throughout the line not only for set-up losses, but also for other availability losses and performance losses as well. The degree of propagation ranges according to the buffer size amid zero (total dependence-maximum propagation) and maximum buffer size (total independence-no propagation). It will be debated in the following (§ 3.5.3), when considering the performance losses, although the same principles can be applied to avoid propagation of minor set up losses (mostly for short set-up/changeover, like adjustment and calibrations).

4.5.2. Maintenance availability

The availability of an equipment [24] is defined as $A_{eff} = \frac{T_u}{T_t}$. The availability of the whole production system can be defined similarly. Nevertheless it depends upon the equipment configurations. Operations Manager, through the choice of equipment configurations can increase the maintenance availability. This is a design decision, since different equipments must be bought and installed according to desired availability level. The choice of the configuration usually results as a trade-off between equipment costs and system availability. The two main equipment configuration (not-redundant system, redundant system) are debated in the following.

Not redundant system

When a system is composed of non redundant equipment, each station produces only if the equipment is working.

Hence if we consider a line of n equipment connected a s a series we have that the downtime of each equipment causes the downtime of the whole system.

$$A_{system} = \prod_{i=1}^{n} A_i \tag{10}$$

$$A_{system} = \prod_{i=1}^{n} A_i = 0, 7*0, 8*0, 9 = 0, 504 \tag{11}$$

The availability of system composed of a series of equipment is always lower than the availability of each equipment (figure 6).

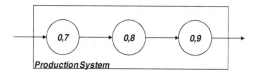

Figure 6. Availability of not redundant System

Total redundant system

Oppositely, to avoid failure propagation amid stations, designer can set the line with a total redundancy of a given equipment. In this case only the contemporaneous downtime of both equipments causes the downtime of the whole system.

$$A_{system} = 1 - \prod_{i=1}^{n} (1 - A_i) \tag{12}$$

In the example in figure 7 we have two single equipments connected with a redundant system of two equipment (dotted line system).

Hence, the redundant system availability (dotted line system) rises from 0,8 (of the single equipment) up to:

$$A_{parallel} = 1 - \prod_{i=1}^{n} (1 - A_i) = \left(1 - 0, 8\right) * \left(1 - 0, 8\right) = 0, 96 \tag{13}$$

Consequently the availability of the whole system will be:

$$A_{system} = \prod_{i=1}^{n} A_i = 0,\ 7^* \left[0,\ 96 \right]^* 0,\ 9 = 0,\ 6048 \tag{14}$$

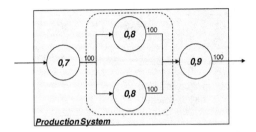

Figure 7. Availability of totally redundant equipments connected with not redundant equipments

To achieve an higher level of availability it has been necessary to buy two identical equipments (double cost). Hence, the higher value of availability of the system should be worth economically.

Partial redundancy

An intermediate solution can be the partial redundancy of an equipment. This is named K/n system, where n is the total number of equipment of the parallel system and k is the minimum number of the n equipment that must work properly to ensure the throughput is produced. The figure 8 shows an example.

The capacity of equipment b', b'' and b''' is 50 pieces in the referral time unit. If the three systems must ensure a throughput of 100 pieces, it is at least necessary that $k=2$ of the $n=3$ equipment produce 50 pieces. The table 4 shows the configuration states which ensure the output is produced and the relative probability that each state manifests.

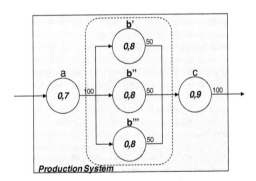

Figure 8. Availability of partially redundant equipments connected with not redundant equipments

b'	b''	b'''	Probability of occurrance	[*100]
UP	UP	UP	0,8*0,8*0,8	0,512
UP	UP	DOWN	0,8*0,8*(1-0,8)	0,128
UP	DOWN	UP	0,8*(1-0,8)*0,8	0,128
DOWN	UP	UP	(1-0,8)*0,8*0,8	0,128
		Total Availability		0,896

Table 4. State Analysis Configuration

In this example all equipments b have the same reliability $(0,8)$, hence the probability the system of three equipment ensure the output should have been calculated, without the state analysis configuration (table 4), through the binomial distribution:

$$R_{k/n} = \sum_{j=k}^{n} \binom{n}{j} R^{j}[1 - R]^{n-j} \tag{15}$$

$$R_{2/3} = \binom{3}{2} 0,8^{2}[1 - 0,8] + \binom{3}{3} 0,8^{3} = 0,896 \tag{16}$$

Hence, the availability of the system (a, b'-b''-b''', c) will be:

$$A_{system} = \prod_{i=1}^{n} A_{i} = 0,7*[0,896]*0,9 = 0,56448 \tag{17}$$

In this case the investment in redundancy is lower than the previous. It is clear how the choice of the level of availability is a trade-off between fix-cost (due to equipment investment) and lack of availability.

In all the cases we considered the buffer as null.

When reliability of the equipments (b in our example) the binomial distribution (16) is not applicable, therefore the state analysis configuration (table 4) is required.

Redundancy with modular capacity

Another configuration is possible.

The production system can be designed as composed of two equipment which singular capacity is lower than the requested but which sum is higher. In this case if it is possible to modulate the production capacity of previous and successive stations the expected throughput will be higher than the output of a singular equipment.

Considering the example in figure 9 when b' and b'' are both up the throughput of the subsystem b'-b'' is 100, since capacity of a and c is 100. Supposing that capacity of a and c is modular, when b' is down the subsystem can produce 60 pieces in the time unit. Similarly, when b'' is down the subsystem can produce 70. Hence, the expected amount of pieces produced by b'-b'' is 84,8 pieces (table 5).

When considering the whole system if either a or c are down the system cannot produce. Hence, the expected throughput in the considered time unit must be reduced of the availability of the two equipments:

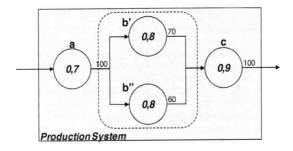

Figure 9. Availability of partially redundant equipments connected with not redundant equipments at modular capacity

b'	b''	Maximum Throughput	Probability of occurrence	[*100]	Expected Pieces Produced
UP	UP	100	0,8*0,8	0,64	64
UP	DOWN	70	0,8*(1-0,8)	0,16	11,2
DOWN	UP	60	(1-0,8)*0,8	0,16	9,6
		Expected number of Pieces Produced			84,8

Table 5. State Analysis Configuration

4.5.3. Minor stoppages and speed reduction

OEE theory includes in performance losses both the cycle time slowdown and minor stoppages. Also time losses of this category propagate, as stated before, throughout the whole production process.

A first type of performance losses propagation is due to the propagation of minor stoppages and reduced speed among machines in series system. From theoretical point of view, between two machines with the same cycle time [1]and without buffer, minor stoppage and reduced speed propagate completely like as major stoppage. Obviously just a little buffer can mitigate the propagation.

Several models to study the role of buffers in avoiding the propagation of performance losses are available in *Buffer Design for Availability* literature [22]. The problem is of scientific relevance, since the lack of opportune buffer between the two stations can indeed affect dramatically the availability of the whole system. To briefly introduce this problem we refer to a production system composed of two consecutive equipments (or stations) with an interposed buffer (figure 10).

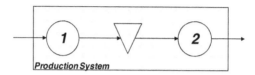

Figure 10. Station-Buffer-Station system. Adapted by [23]

Under the likely hypothesis that the ideal cycle times of the two stations are identical [23], the variability of speed that affect the stations is not necessarily of the same magnitude, due to its dependence on several factors. Furthermore Performance index is an average of the T_t , therefore a same machine can sometimes perform at a reduced speed and sometimes an highest speed[2]. The presence of this effect in two consecutive equipments can be mutually compensate or add up. Once again, within the propagation analysis for production system design, the role of buffer is dramatically important.

When buffer size is null the system is in series. Hence, as for availability, speed losses of each equipment affect the performance of the whole system:

$$P_{system} = \prod_{i=1}^{n} P_i \tag{18}$$

Therefore, for the two stations system we can posit:

---

1 As shown in par. 3.1. When two consecutive stations present different cycle times, the faster station works with the same cycle time of slower station, with consequence on equipment OEE, even if any time losses is occurred. On the other hand, when two consecutive stations are balanced (same cycle time) if any time loss is occurring the two stations OEE will be 100%. Ideally, the higher value of performance rate can be reached when the two stations are balanced.

2 This time losses are typically caused by yield reduction (the actual process yield is lower than the design yield). This effect is more likely to be considered in production process where the equipment saturation level affect its yield, like furnaces, chemical reactor, etc.

$$P_{system} = \prod_{i=1}^{2} P_i \qquad (19)$$

But when the buffer is properly designed, it doesn't allow the minor stoppages and speed losses to propagate from a station to another. We define this Buffer size as Bmax. When, in a production system of n stations, given any couple of consecutive station, the interposed buffer size is Bmax (calculated on the two specific couple of stations), then we have:

$$P_{system} = Min_{i=1}^{n}(P_i) \qquad (20)$$

That for the considered 2 stations system is:

$$P_{system} = Min\ (P_1,\ P_2) \qquad (21)$$

Hence, the extent of the propagation of performance losses depends on the buffer size (j) that is interposed between the two stations. Generally, a bigger buffer increases the performance of the system, since it increases the decoupling degree between two consecutive stations, up to j=Bmax is achieved (j =0,..,Bmax).

We can therefore introduce the parameter

$$\mathbf{Rel.P}\left(\mathbf{j}\right) = \frac{P\ (j)}{P(Bmax)} \qquad (22)$$

Considering the model with two station, figure 11, we have that:

$$\mathbf{When\ j}\ =\ 0,\ \ \mathbf{Rel.P}\left(0\right) = \frac{P\ (0)}{P(Bmax)} = P\left(1\right)*P\left(2\right)/\min\left(P\left(1\right); P\left(2\right)\right); \qquad (23)$$

$$\mathbf{When\ j}\ =\ \mathbf{Bmax},\ \ \mathbf{Rel.P}\left(\mathbf{B\ max}\right) = \frac{P\ (Bmax)}{P(Bmax)} = 1; \qquad (24)$$

Figure 11 shows the trend of *Rel.P(j)* depending on the buffer size (*j*), when the performance rate of each station is modeled with an exponential distribution [23] in a flow shop environment. The two curves represent the minimum and the maximum simulation results. All the others simulation results are included between these two curves. Maximum curve represents the configuration with the lowest difference in performance index between the two stations, the minimum the configuration with the highest difference.

By analyzing the figure 11 it is clear how an inopportune buffer size affect the performance of the line and how increase in buffer size allows to obtain improve in production line OEE. By the way, once achieved an opportune buffer size no improvement derives from a further increase in buffer. These considerations of Performance index trend are fundamental for an effective design of a production system.

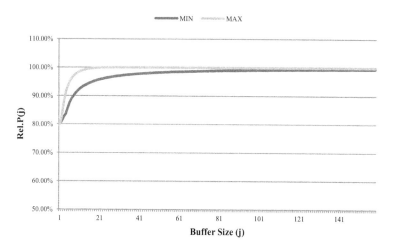

Figure 11. Rel OEE depending on buffer size in system affected by variability due to speed losses

4.5.4. Quality losses

In this paragraph we analyze how quality losses propagate in the system and if it is possible to assess the effect of quality control on OEE and OTE.

First of all we have to consider that quality rate for a station is usually calculated considering only the time spent for the manufacturing of products that have been rejected in the same station. This traditional approach focuses on stations that cause defects but doesn't allow to point out completely the effect of the machine defectiveness on the system. In order to do so, the total time wasted by a station due to quality losses should include even the time spent for manufacturing of good products that will be rejected for defectiveness caused by other stations. In this sense quality losses depends on where scraps are identified and rejected. For example, scraps in the last station should be considered loss of time for the upstream station to estimate the real impact of the loss on the system and to estimate the theoretical production capacity needed in the upstream station. In conclusion the authors propose to calculate quality rate for a station considering as quality loss all time spent to manufacture products that will not complete the whole process successfully.

From a theoretical point of view we could consider the following case for calculation of quality rate of a station that depends on types of rejection (scraps or rework) and on quality controls positioning. If we consider two stations with an assigned defectiveness S_j and each station reworks its scraps with a rework cycle time equal to theoretical cycle time, quality rate could be formulate as shown in case 1 in figure 12. Each station will have quality losses (time spent to rework products) due its own defectiveness. If we consider two stations with an assigned defectiveness S_j and a quality control station at downstream each station, quality rate could be formulate as shown in case 2 in figure 12. The station 1, that is the upstream station, will have quality losses (time spent to work products that will be discarded) due to its own and station 2 defectiveness. If we consider two stations with an assigned defectiveness S_j and quality control station is only at the end of the line, quality rate quality rate could be formulate as shown in case 3 in figure 12. In this case both stations will have quality losses due to the propagation of defectiveness in the line. Case 2 and 3 point out that quality losses could be not simple to evaluate if we consider a long process both in design and management of system. In particular in the quality rate of station 1 we consider time lost for reject in the station 2.

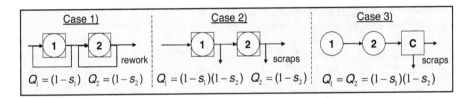

Figure 12. Different cases of quality rate calculation

Finally, it is important to highlight the different role that the quality efficiency plays during the design phase and the production.

When the system is producing, Operations Manager focuses his attention on the causes of the delectability with the aim to reduce it. When it is to design the production system, Operations Manager focuses on the expected quality efficiency of each station, on the location of quality control, on the process (rework or scraps) to identify the correct number of equipments or station for each activity of the process.

In this sense, the analysis is vertical during the production phase, but it follows the whole process during the design (figure 13).

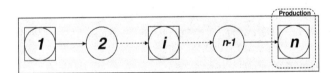

Figure 13. Two approaches for quality efficiency

5. The simulation model

To study losses propagation and to show how these dynamics affect OEE in a complex system [25] this chapter presents some examples taken from an OEE study of a real manufacturing system carried out by the authors through a process simulation analysis [19].

Simulation is run for each kind of time losses (Availability, Performance and Quality), to clearly show how each equipment ineffectiveness may compromise the performance of the whole system.

The simulation model is about a manufacturing plant for production of electrical cable. In particular we focuses on production of unipolar electrical cable that takes place by a flow-shop process. In the floor plant the production equipment is grouped in production areas arranged according to their functions (process layout). The different production areas are located along the line of product flow (product layout). Buffers are present amongst the production areas to stock the product in process. This particular plant allows to analyze deeply the problem of OEE-OTE investigation due to its complexity.

In terms of layout the production system was realized as a job shop system, although the flow of material from a station to another was continuous and typical of flow shop process. As stated in (§2) the reason lies on due to the huge size of the products that passes from a station to another. For this reason the buffer amid station, although present, couldn't contain huge amount of material.

The process implemented in the simulation model is shown in figure 14. Entities are unit quantity of cable that have different mass amongst stations. Parameters that are data input in the model are equipment speed, defectiveness, equipment failure rate and mean time to repair. Each parameters is described by a statistical distribution in order to simulate random condition. In particular equipment speed has been simulated with a triangular distribution in order to simulate performance losses due to speed reduction.

The model evaluates OTE and OEE for each station as usually measured in manufacturing plant. The model has been validated through a plan of tests and its results of OEE has been compared with results obtained from an analytic evaluation.

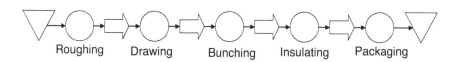

Roughing   Drawing   Bunching   Insulating   Packaging

Figure 14. ASME representation of manufacturing process

5.1. Example of availability losses propagation

In accordance with the proposed method (§ 3.5) we show how availability losses propagate in the system and to assess the effect of buffer capacity on OEE through the simulation. We focuses on the insulating and packaging working stations. Technical data about availability of equipment are: mean time between failure for insulating is 20000 sec while for packaging is 30000 sec; mean time between repair for insulating is 10000 sec while for packaging is 30000 sec. The cycle time of the working stations are the same equal to 2800 sec for coil. The quality rates are set to 1. Idling, minor stoppages and reduced speed are not considered and set to 0.

Considering equipment isolated from the system the OEE for the single machine is equal to its availability; in particular, relating to previous data, machines have an OEE equal to 0,67 and 0,5 respectively for insulating and packaging. The case points out how the losses due to major stoppage spread to other station in function of buffer capacity dimension.

A simulation has been run to study the effect of buffer capacity in this case. Capacity of buffer downstream of insulating has been changed from 0 to 30 coils for different simulations. The results of simulations are shown in figure 15a. The OEE for both machines is equal to 0,33 with no buffer capacity. This results is the composition of availability of insulating and packaging (0,67 x 0,5) as expected. The OEEs increase in function of buffer dimension that avoids the propagation of major stoppage and availability losses propagation. Also the OTE is equal to 0,33 that is, according to formulation in (1) and as previously explained, equal to OEE of the last station but assessed in the system.

Insulating and packaging increase rapidly OEEs since a structural limits of buffer capacity of 15 coils; from this value OEEs of two stations converge on value of 0,5. The upstream insulating station, that has an availability greater than packaging, has to adapt itself to real cycle time of packaging that is the bottleneck station.

It's important to point out that in performance monitoring of manufacturing plant the propagation of the previous losses is often gathered as performance losses (reduced speed or minor stoppage) in absence of specific data collection relating to major stoppage due to absence of material flow. So, if we consider also all other efficiency looses ignored in this sample, we can understand how much could be difficult to identify the real impact of this kind of efficiency losses monitoring the real system. Moreover simulation supports in system design in order to dimension buffer capacity (e.g. in this case structural limit for OEE is reached for 16 coils). Moreover through simulation it is possible to point out that the positive effect of buffer is reduced with an higher cycle time of machine as shown in figure 15b.

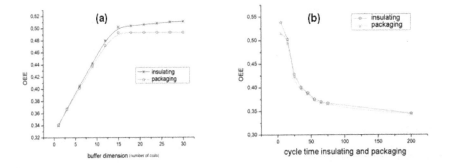

Figure 15. OEE in function of buffer dimension (a) and cycle time (b)

5.2. Minor stoppages and speed reduction

We run the simulation also for the case study (§ 4). The simulation shown how two stations, with the same theoretical cycle time (200 sec/coil) affected by a triangular distribution with a performance rate of 52% as single machine, have: 48% of performance rate with a capacity buffer of 1 coil and 50% of performance rate with a capacity buffer of 2 coils. But if we consider two stations with the same theoretic cycle time but affects by different triangular distributions so that theoretic performance rates differ, simulation shows how the performance rates of two stations converge towards the lowest one as expected (19), (20).

Through the same simulation model we considered also the second type of performance losses propagation, due to the propagation of reduced speed caused by unbalanced line. Figure 16 shown the effect of unbalanced cycle time of stations relating to insulating and packaging. The station have the same P as single machine equal to 67% but different theoretical cycle time. In particular insulating, the upstream station, is faster than packaging. Availability and quality rate of stations is set to 1. The buffer capacity is set to 1 coil. A simulation has been run to study the effect of unbalancing station. Theoretical cycle time of insulating has been changed since theoretical cycle time of packaging that is fixed in mean. The simulation points out that insulating has to adapt itself to cycle time of packaging that is the bottleneck station. This results in the model as a lower value for performance rate of insulating station. The same happens often in real systems where the result is influenced by all the efficiency losses at the same time. The effect disappears gradually with a better balancing of two stations as in figure 16.

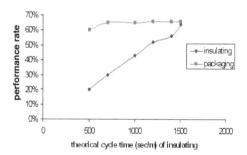

Figure 16. Performance rate of insulating and packaging in function of insulating cycle time

5.3. Quality losses

In relation to the model, this sample focuses on the drawing and bunching working stations that have defectiveness set to 5%, the same cycle times and no other efficiency losses. The quality control has been changed simulating case 2 and 3. The results of simulation for the two cases are shown in table 6 in which the proposal method has compared with the traditional one. The proposal method allowed to identify the correct efficiency, for example to dimension the drawing station, because it considers time wasted to manufacture products rejected in bunching station. The difference between values of Q_2 and OTE is explained by the value of $P_2=0,95$ that is due to the propagation of quality losses for the upstream station in performance losses for the downstream station. Moreover about positioning of quality control the case 2 has to be prefer because the simulation shows a positive effect on the OTE if the bunching station is the system bottleneck (as it happens in the real system).

	Proposal method			Traditional method		
	Q1	Q2	OTE	Q1	Q2	OTE
Case 2)	$0,95^2$	0,95	$0,95^2$	0,95	0,95	$0,95^2$
Case 3)	$0,95^2$	$0,95^2$	$0,95^2$	--	$0,95^2$	$0,95^2$

Table 6. Comparison of quality rate calculation and evaluation of impact of quality control positioning on quality rates and on OTE

6. Conclusions

The evaluation of Overall Equipment Effectiveness (OEE) and Overall Throughput Effectiveness (OTE) can be critical for the correct estimation of workstations number needed to realize the desired throughput (production system design), as also for the analysis and the continuous improvement of the system performance (during the system management).

The use of OEE as performance improvement tool has been widely described in the literature. But it has been less approached in system design for a correct evaluation of the system efficiency (OTE), in order to study losses propagation, overlapping of efficiency losses and effective actions for losses reduction.

In this chapter, starting by the available literature on time losses, we identified a simplified set of relevant time-losses that need to be considered during the design phase. Then, through the simulation, we shown how OEE of single machine and the value of OTE of the whole system are interconnected and mutually influencing each other, due to the propagation of availability, performance and quality losses throughout the system.

For each category of time losses we described the effects of efficiency losses propagation from a station to the system, for a correct estimation and analysis of OEE and OTE during manufacturing system design. We also shown how to avoid losses propagation through adequate technical solutions which can be defined during system design as the buffer sizing, the equipment configuration and the positioning of control stations.

The simulation model shown in this chapter was based on a real production system and it used real data to study the losses propagation in a manufacturing plant for production of electrical cable. The validation of the model ensures the meaningful of the approach and of the identified set of possible solutions and hints.

By analyzing and each time losses we also shown how the choices taken during the design of the production system to increase the OTE (e.g. buffer size, maintenance configuration, etc.) affect the successive management of the operations.

Acknowledgements

The realization of this chapter would not have been possible without the support of a person whose cooperated with the chair of Operations Management of University of Rome "Tor Vergata" in the last years, producing valuable research. The authors wish to express their gratitude to Dr. Bruna Di Silvio without whose knowledge, diligence and assistance this work would not have been successful.

Author details

Vittorio Cesarotti[1], Alessio Giuiusa[1,2] and Vito Introna[1]

1 University of Rome "Tor Vergata", Italy

2 Area Manager Inbound Operations at Amazon.com

References

[1] H. H. S., «Manufacturing productivity improvement using effectivenes metrics and simulation analysis,» 2002.

[2] B. I., «Effective measurement and successful elements of company productivity: the basis of competitiveness and world prosperity,» *International Journal of Production Economics*, vol. 52, pp. 203-213, 1997.

[3] Jeong, P. D. K.Y., «Operational efficiency and effectiveness measurement,» *International Journal of Operations and Production Management*, n. 1404-1416, (2001). , 21

[4] N. S., Introduction to TPM- Total Productive Maintenance, Productivity Press, 1988.

[5] S. R.J., World Class Manufacturing. The lesson of simplicity Applied, The Free Press, 1987.

[6] Womack, J. D. J.P., Lean Thinking, Simon & Schuster, (1996).

[7] Dixon, N. A. V. J.R., The new performance challenge. Measuring operations for world-class competition, Dow Jones Irwin, (1990).

[8] S. D., «Can CIM improve overall factory effetivenes,» in *Pan Pacific Microelectronic Symposium*, Kauai, HI, 1999.

[9] Waters, W. D. D.J., Operations Management, Kogan Page Publishers, (1999).

[10] Chase, A. N. J. F. R.B., Operations Management, McGraw-Hill, (2008).

[11] A. V. A., Semiconductor Manufacturing Productivity- Overall Equipment Effectiveness (OEE) guidebook, SEMATECH, 1995.

[12] Rooda, D. R. A. J. J.E., «Equipment effectiveness: OEE revisited,» *IEEE Transactions on Semiconductor Manufacturing*, n. 1, (2005). , 18 , 189-196.

[13] Gamberini, G. L. R. B. R., «Alternative approaches for OEE evaluation: some guidelines directing the choice,» in *XVII Summer School Francesco Turco*, Venice, (2012).

[14] Grando, T. F. A., «Modelling Plant Capacity and Productivity,» *Production Planning and Control*, n. 3, (2005). , 16 , 209-322.

[15] Spada, C. V. C., «The Impact of Cultural Issues and Interpersonal Behavior on Sustainable Excellence and Competitiveness: An Analysis of the Italian Context,» *Contributions to Management Science*, (2008). , 95 - 113 .

[16] Badiger, G. R, & Proposal, A. , «A. evaluation of OEE and impact of six big losses on equipment earning capacity,» *International Journal Process Management & Benchmarking*, (2008). , 235 - 247 .

[17] Jacobs, A. N. C. B, & Operations, R. F. and supply chain management, McGraw-Hill, A cura di, (2010).

[18] Oechsner, R. From overall equipment efficiency(OEE) to overall Fab effectiveness (OFE),» *Materials Science in Semiconductor Processing*, (2003). , 5 , 333-339.

[19] Introna, D. S. B. C. V, & Flow-shop, V. process oee calculation and improvement using simulation analysis,» in *MITIP*, Florence, (2007).

[20] R. MA, «Factory Level Metrics: Basis for Productivity Improvement,» in *Proceedings of the International Conference on Modeling and Analysis of Semiconductor*, Tempe, Arizona, USA, 2002.

[21] Scott, P. R. D., «Can overall factory effetiveness prolong Moore's Law?,» *Solid State Technology*, (1998). , 41 , 75-82.

[22] B. D., «Buffer size design linked to reliability performance: A simulative study,» *Computers & Industrial Engineering*, vol. 56, p. 1633-1641, 2009.

[23] Introna, G. A. V. V., «Increasing Availability of Production Flow lines through Optimal Buffer Sizing: a Simulative Study,» in *The 23rd European Modeling & Simulation Symposium (Simulation in Industry)*, Rome, (2011).

[24] Connor, P. D. T. O. Practical Reliability Engineering (Fourth Ed.), New York: John Wiley & Sons, (2002).

[25] Kane, J. O. Simulating production performance: cross case analysis and policy implications,» *Industrial Management & Data Systems,* n. 4, (2004). , 104 , 309-321.

[26] Gondhinathan, B. A, & Proposal, R. , «A. evaluation of OEE and impact of six big losses on equipment earning capacity,» *International Journal Process Management & Benchmarking,* n. 3, (2008). , 2 , 235-247.

Chapter 3

Managing OEE to Optimize Factory Performance

Raffaele Iannone and Maria Elena Nenni

Additional information is available at the end of the chapter

1. Introduction

"If you can not measure it, you can not improve it."(Lord Kelvin)

It is a common opinion that productivity improvement is nowadays the biggest challenge for companies in order to remain competitive in a global market [1, 2]. A well-known way of measuring the effectiveness is the Overall Equipment Efficiency (OEE) index. It has been firstly developed by the Japan Institute for Plant Maintenance (JIPM) and it is widely used in many industries. Moreover it is the backbone of methodologies for quality improvement as TQM and Lean Production.

The strength of the OEE index is in making losses more transparent and in highlighting areas of improvement. OEE is often seen as a catalyst for change and it is easy to understand as a lot of articles and discussion have been generated about this topic over the last years.

The aim of this chapter is to answer to general questions as *what to measure? how to measure?* and *how to use the measurements?* in order to optimize the factory performance. The goal is to show as OEE is a good base for optimizing the factory performance. Moreover OEE's evolutions are the perfect response even in advanced frameworks.

This chapter begins with an explanation of the difference between efficiency, effectiveness and productivity as well as with a formal definition for the components of effectiveness. Mathematical formulas for calculating OEE are provided too.

After the introduction to the fundamental of OEE, some interesting issues concerning the way to implement the index are investigated. Starting with the question that in calculating OEE you have to take into consideration machines as operating in a linked and complex environment. So we analyze almost a model for the OEE calculation that lets a wider approach to the performance of the whole factory. The second issue concerns with monitoring the factory performance through OEE. It implies that information for decision-making have to be guaranteed real-time. It is possible only through automated systems for calculating OEE and through the capability to collect a large amount of data. So we propose an examination of the main automated OEE systems from the simplest to high-level systems integrated into ERP software. Even data collection strategies are screened for rigorous measurement of OEE.

The last issue deals with how OEE has evolved into tools like TEEP, PEE, OFE, OPE and OAE in order to fit with different requirements.

At the end of the chapter, industrial examples of OEE application are presented and the results are discussed.

2. Fundamentals of OEE

Overall equipment efficiency or effectiveness (OEE) is a hierarchy of metrics proposed by Seiichi Nakajima [3] to measure the performance of the equipment in a factory. OEE is a really powerful tool that can be used also to perform diagnostics as well as to compare production units in differing industries. The OEE has born as the backbone of Total Productive Mainte-nance (TPM) and then of other techniques employed in asset management programs, Lean manufacturing [4], Six Sigma [5], World Class Manufacturing [4].

By the end of the 1980's, the concept of Total Production Maintenance became more widely known in the Western world [7] and along with it OEE implementation too. From then on an extensive literature [8-11] made OEE accessible and feasible for many Western companies.

3. Difference between efficiency, effectiveness and productivity

Confusion exists as to whether OEE has indeed been an effectiveness or efficiency measure. The traditional vision of TMP referred to Overall Equipment Efficiency while now it is generally recognized as Overall Equipment Effectiveness. The difference between efficiency and effectiveness is that effectiveness is the actual output over the reference output and efficiency is the actual input over the reference input. The Equipment Efficiency refers thus to ability to perform well at the lowest overall cost. Equipment Efficiency is then unlinked from output and company goals. Hence the concept of Equipment Effectiveness relates to the ability of producing repeatedly what is intended producing, that is to say to produce value for the company (see Figure 1).

Productivity is defined as the actual output over the actual input (e.g. number of final products per employee), and both the effectiveness and the efficiency can influence it. Regarding to OEE, in a modern, customer-driven "lean" environment it is more useful to cope with effectiveness.

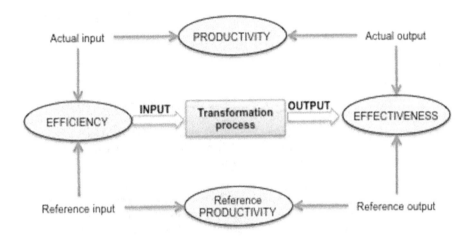

Figure 1. Efficiency versus Effectiveness versus Productivity.

4. Formal definition of OEE

According to the previous remark a basic definition of OEE is:

$$OEE = \frac{ValuableOperatingTime}{LoadingTime} \tag{1}$$

where:

- Valuable Operating Time is the net time during which the equipment actually produces an acceptable product;

- Loading Time is the actual number of hours that the equipment is expected to work in a specific period (year, month, week, or day).

The formula indicates how much the equipment is doing what it is supposed to do and it captures the degree of conforming to output requirements. It is clearly a measure of effectiveness.

OEE is not only a metric, but it also provides a framework to improve the process. A model for OEE calculation aims to point out each aspect of the process that can be ranked for improvement. To maximize equipment effectiveness it is necessary to bring the equipment to peak operating conditions and then keeping it there by eliminating or at least minimizing any factor that might diminish its performance. In other words a model for OEE calculation should be based on the identification of any losses that prevent equipment from achieving its maximum effectiveness.

The OEE calculation model is then designed to isolate losses that degrade the equipment effectiveness.

5. Losses analysis

Losses are activities that absorb resources without creating value. Losses can be divided by their frequency of occurrence, their cause and by different types they are. The latter one has been developed by Nakajima [3] and it is the well-known Six Big Losses framework. The other ones are interesting in order to rank rightly losses.

According to Johnson et al. [12], losses can be chronic or sporadic. The chronic disturbances are usually described as *"small, hidden and complicated"* while the sporadic ones occur quickly and with large deviations from the normal value. The loss frequency combined with the loss severity gives a measure of the damage and it is useful in order to establish the order in which the losses have to be removed. This classification makes it possible to rank the losses and remove them on the basis of their seriousness or impact on the organization.

Regarding divide losses by their causes, three different ones can be found:

1. machine malfunctioning: an equipment or a part of this does not fulfill the demands;

2. process: the way the equipment is used during production;

3. external: cause of losses that cannot be improved by the maintenance or production team.

The external causes such as shortage of raw materials, lack of personnel or limited demand do not touch the equipment effectiveness. They are of great importance for top management and they should be examined carefully because their reduction can directly increase the revenues and profit. However they are not responsible of the production or maintenance team and so they are not taken into consideration through the OEE metric.

To improve the equipment effectiveness the losses because of external causes have to be taken out and the losses caused by machine malfunctioning and process, changeable by the daily organization, can still be divided into:

*   **Down time losses:** when the machine should run, but it stands still. Most common downtime losses happen when a malfunction arises, an unplanned maintenance task must be done in addition to the big revisions or a set-up/start-up time occurs.

*   **Speed losses:** the equipment is running, but it is not running at its maximum designed speed. Most common speed losses happen when equipment speed decrease but it is not zero. It can depend on a malfunctioning, a small technical imperfections, like stuck packaging or because of the start-up of the equipment related to a maintenance task, a setup or a stop for organizational reasons.

*   **Quality losses:** the equipment is producing products that do not fully meet the specified quality requirements. Most common quality losses occur because equipment, in the time

between start-up and completely stable throughput, yields products that do not conform to quality demand or not completely. They even happen because an incorrect functioning of the machine or because process parameters are not tuned to standard.

The framework in which we have divided losses in down time, speed and quality losses completely fits with the Six Big Losses model proposed by Nakajima [3] and that we summarize in the Table 1:

Category	Big losses
DOWNTIME	- Breakdown - Set-up and adjustments
SPEED	- Idling, minor stoppages - Reduced speed
QUALITY	- Quality losses - Reduced yield

Table 1. Six Big Losses model proposed by Nakajima [3].

On the base of Six Big Losses model, it is possible to understand how the Loading Time decreases until to the Valuable Operating Time and the effectiveness is compromised. Let's go through the next Figure 2.

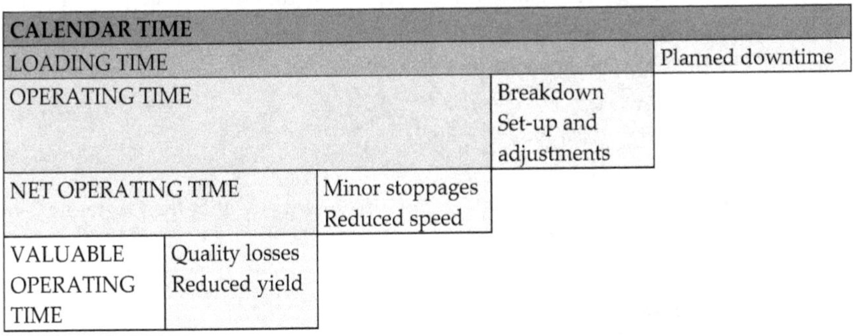

CALENDAR TIME			
LOADING TIME			Planned downtime
OPERATING TIME		Breakdown Set-up and adjustments	
NET OPERATING TIME		Minor stoppages Reduced speed	
VALUABLE OPERATING TIME	Quality losses Reduced yield		

Figure 2. Breakdown of the calendar time.

At this point we can define:

$$Availability\ (A) = \frac{Operating\ Time}{Loading\ Time} \qquad (2)$$

$$Performance\ (P) = \frac{Net\ Operating\ Time}{Operating\ Time} \qquad (3)$$

$$Quality\ (Q) = \frac{Valuable\ Operating\ Time}{Net\ Operating\ Time} \qquad (4)$$

Please note that:

$$OEE = \frac{Valuable\ Operating\ Time}{Loading\ Time} \qquad (5)$$

and

$$OEE = \frac{Operating\ Time}{Loading\ Time} \times \frac{Net\ Operating\ Time}{Operating\ Time} \times \frac{Valuable\ Operating\ Time}{Net\ Operating\ Time} \qquad (6)$$

finally

$$OEE = Availability \times Performance \times Quality \qquad (7)$$

So through a bottom-up approach based on the Six Big Losses model, OEE breaks the performance of equipment into three separate and measurable components: Availability, Performance and Quality.

- **Availability:** it is the percentage of time that equipment is available to run during the total possible Loading Time. Availability is different than Utilization. Availability only includes the time the machine was scheduled, planned, or assigned to run. Utilization regards all hours of the calendar time. Utilization is more effective in capacity planning and analyzing fixed cost absorption. Availability looks at the equipment itself and focuses more on variable cost absorption. Availability can be even calculated as:

$$Availability = \frac{Loading\ Time - Downtime}{Loading\ Time} \qquad (8)$$

- **Performance:** it is a measure of how well the machine runs within the Operating Time. Performance can be even calculated as:

$$Performance = \frac{Actual\ Output\ (units) \times theoretical\ Cycle\ Time}{Operating\ Time} \qquad (9)$$

- **Quality:** it is a measure of the number of parts that meet specification compared to how many were produced. Quality can be even calculated as:

$$Quality = \frac{Actual\ output\ (units) - Defect\ amount\ (units)}{Actual\ output\ (units)} \qquad (10)$$

After the various factors are taken into account, all the results are expressed as a percentage that can be viewed as a snapshot of the current equipment effectiveness.

The value of the OEE is an indication of the size of the technical losses (machine malfunctioning and process) as a whole. The gap between the value of the OEE and 100% indicates the share of technical losses compared to the Loading Time.

The compound effect of Availability, Performance and Quality provides surprising results, as visualized by e.g. Louglin [13].

Let's go through a practical example in the Table 2.

Availability	86,7%
Performance	93%
Quality	95%
OEE	76,6%

Table 2. Example of OEE calculation.

The example in Table 2 illustrates the sensitivity of the OEE measure to a low and combined performance. Consequently, it is impossible to reach 100 % OEE within an industrial context. Worldwide studies indicate that the average OEE rate in manufacturing plants is 60%. As pointed out by e.g. Bicheno [14] world class level of OEE is in the range of 85 % to 92 % for non-process industry. Clearly, there is room for improvement in most manufacturing plants! The challenge is, however, not to peak on those levels but thus to exhibit a stable OEE at world-class level [15].

6. Attacking the six big losses

By having a structured framework based on the Six Big Losses, OEE lets to track underlying issues and root causes. By knowing what the Six Big Losses are and some of the causes that contribute to them, the next step is to focus on ways to monitor and correct them. In the following let's see what is the way:

- **Breakdown:** eliminating unplanned downtime is critical to improving OEE. Other OEE factors cannot be addressed if the process is down. It is not only important to know how much and when down time equipment is but also to be able to link the lost time to the specific source or reason for the loss. With down time data tabulated, the most common approach is the Root Cause Analysis. It is applied starting with the most severe loss categories.

- **Set-up and adjustments:** tracking setup time is critical to reducing this loss. The most common approach to reduce this time is the Single Minute Exchange of Dies program (SMED).

- **Minor stoppages and Reduced speed:** minor stoppages and reduced speed are the most difficult of the Six Big Losses to monitor and record. Cycle Time analysis should be utilized to point out these loss types. In most processes recording data for Cycle Time analysis needs to be automated since the cycles are as quick as they do not leave adequate time for manual data logging. By comparing all cycles to the theoretical Cycle Time, the losses can be automatically clustered for analysis. It is important to analyze Minor stoppages and Reduced speed separately because the root causes are typically very different.

- **Quality losses and Reduced yield:** parts that require rework of any kind should be considered rejects. Tracking when rejects occur and the type is critical to point out potential causes, and in many cases patterns will be discovered. Often a Six Sigma program, where a common metric is achieving a defect rate of less than 3.4 defects per million opportunities, is used to focus attention on a goal of "zero defects".

7. OEE evolution: TEEP, PEE, OAE, OFE, and OPE

During the last decades, both practitioners and researchers have raised the discussion about OEE in many ways. One of the most popular has led to modification and enlargement of individual original OEE tool to fit a broader perspective as supposed important for the companies [16]. With the evolution of OEE, different definitions have also come up in literature and in practice, coupled with their changed formulations. Some of these formulations (TEEP and PEE) are still at the equipment level, while the others (OAE, OFE and OPE) extended OEE to the factory level. Let's go through the main features of each formulation.

TEEP stands for **Total Equipment Effectiveness Performance** and it was proposed firstly by Invancic [17]. TEEP is a performance metric that shows the total performance of equipment based on the amount of time the equipment was present. So OEE quantifies how well a manufacturing unit performs relative to its designed capacity, during the periods when it is scheduled to run. TEEP measures OEE effectiveness against Calendar Time, i.e.: 24 hours per day, 365 days per year.

$$TEEP = \frac{Valuable\ Operating\ Time}{Calendar\ Time} = OEE \times \frac{Loading\ Time}{Calendar\ Time} \tag{11}$$

OEE and TEEP are thus two closely related measurements. Typically the equipment is on site and thus TEEP is metric that shows how well equipment is utilized. TEEP is useful for business analysis and important to maximize before spending capital dollars for more capacity.

PEE stands for **Production Equipment Efficiency** and it was firstly proposed by Raouf [18]. The main difference from OEE is that each item is weighted. So Availability, Performance, and Quality don't have an equal importance as it happens for OEE.

At the level of the factory we found **Overall Factory Effectiveness** (OFE), **Overall Production Effectiveness** (OPE), and **Overall Asset Effectiveness** (OAE) metrics. OFE is the most widespread and well known in literature. It covers the effort to export the OEE tool to the whole factory. The question is what kind of method should be applied to OEE values from all pieces of equipment, to derive the factory level metric. There is no standard method or metrics for the measurement or analysis of OFE [19]. Huang [20] stated that the factory level metric can be computed by synthesizing the subsystem level metrics, capturing their interconnectivity information.

OPE and OAE are extensively implemented in industry under different formulations. They involve a practical approach developed to fit the specific requirements of different industries.

8. OEE for the factory

As mentioned in the previous section equipment operates in a linked and complex environment. So it is necessary to pay attention beyond the performance of individual tools towards the performance of the whole factory. According to Scott and Pisa [21], the answer to this requirement is the OFE metric, which is about combining activities and relationships between different parts of the equipment, and integrating information, decisions, and actions across many independent systems and subsystems. The problem is that a specific and unique method to calculate OFE does not exist. There many methodologies and approaches, with different level of complexity, different information coming from and different lacks.

A first common-sense approach is to measure OEE at the end of the line or process. Following this approach we can see OEE as

$$OEE = \frac{(Actual\ output - Defect\ amount) \times theoretical\ Cicle\ Time}{Loading\ Time} \tag{12}$$

and

$$OEE = \frac{Effective\ output\ (units)}{theoretical\ output\ (units)} \tag{13}$$

Here OEE measures effectiveness in term of output that is easy to be taken out at factory level too. So OFE becomes:

$$OFE = \frac{Effective\ output\ from\ the\ factory\ (units)}{Theoretical\ output\ from\ the\ factory\ (units)} \tag{14}$$

It is not always ideal. The complexity of OEE measurement arises where single or multiple sub-cells are constrained by an upstream or downstream operation or bottleneck operation. The flow is always restricted or limited by a bottleneck operation, just as a chain is only as strong as its weakest link. So according to Goldratt [22] we can measure OEE in real time at the bottleneck. Any variations at the bottleneck correlate directly to upstream and downstream process performance. Huang et al. [23] proposed a manufacturing system modeling approach, which captures the equipment interconnectivity information. It identifies four unique subsystems (series, parallel, assembly and expansion) as a basis for modeling a manufacturing system, as shown in Figure 3.

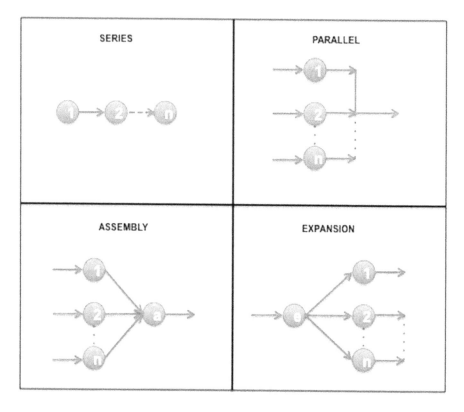

Figure 3. Types of manufacturing subsystems.

Muthiah et al. [24] developed the approach to derive OTE metrics for these subsystems based on a "system constraint" approach that automatically takes into account equipment idle time.

Other methods are based on modeling the manufacturing systems. Some of these notable approaches are queuing analysis methods [25], Markovian methods [26], Petri net based methods [27], integrated computer-aided manufacturing definition (IDEF) method [28], and structured analysis and design technique (SADT) [29]. In addition to them there are several commercial tools that have been reviewed and categorized by Muthiah and Huang [30].

9. What is OEE for?

OEE provides simple and consolidated formulas to measure effectiveness of the equipment or production system. Moreover Dal et al. [31] point out that it can also be used as an indicator of process improvement activities since OEE is directly linked to the losses as well as OEE can be even used to compare performance across the factory highlighting poor line performance or to quantify improvements made [31]. Moreover improving can be pursued by:

• Backtracking to determine what loss reduces effectiveness.

• Identifying bottlenecks as not only the slowest machine, but as the machine both slower and less effective.

All these goals need of an approach based on the Deming Cycle [32]. It is an improvement cycle to increase the plant OEE rating until the target goals and world class manufacturing status are achieved (Figure 4)

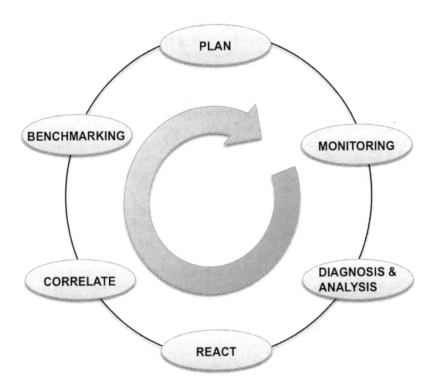

Figure 4. Improvement approach to increase the plant OEE.

This approach requires a large amount of data that can be provided both in a static or dynamic way. In the first case data are picked up only at the end of a certain period and used in the Diagnosis & Analysis stage.

There is another way to use OEE and it is to know exactly what is happening in real time through a continuous monitoring to immediately identify possible problems and react in real-time using appropriate corrective actions. Information on OEE items (maintenance and operational equipment effectiveness, product data accuracy, uptimes, utilization, bottlenecks, yield and scrap metrics, etc.) is really valuable in environments where making decisions in near real-time is critical. This second approach requires then a data collection system completely automatized and moreover the Diagnosis & Analysis stage should be automatic.

In the next sections we will take into consideration different strategies to acquire data and we will illustrate main automated tool for the OEE integration.

10. Data collection strategies

The OEE calculations should be based on correct input parameters from the production system as reported by Ericsson [33]. Data acquisition strategies range from very manual to very automated. The manual data collection method consists of a paper template, where the operators fill in the cause and duration of a breakdown and provide comments about minor stoppages and speed losses. It is a low-tech approach. On the contrary a high-tech approach runs through an automatic OEE calculation system that is governed by sensors connected to the equipment, automatically registering the start time and duration of a stoppage and prompting the operator to provide the system with information about the downtime cause. An automatic approach usually provides opportunities to set up lists of downtime causes, scheduling the available operating time and making an automatic OEE calculation for a time period. A variety of reports of production performance and visualization of the performance results are even possible to retrieve from the system.

Two approaches have to be compared through opportunity and cost both, in a quantitative as well as in a qualitative way. Regarding cost, the main figures in case of the manual approach are derived from the hourly wage cost of operators multiplied by time spent to register data on paper templates, feed them into a computer system and for generating reports and performing OEE calculations. In case of the automatic approach cost concerns a yearly license cost for an automatic OEE calculation system together with an investment cost for hardware. The introduction of both the manual and automatic data collection methods must be preceded and then associated with training of the operators on OEE as a performance measure, and on different parameters affecting the OEE outcome. The purpose of training the operators was twofold:

1. The quality of the input data is likely to increase in alignment with an increase in the competence of the staff;

2. The involvement of the operators in identifying performance loss factors is likely to create a better engagement for providing the system with accurate information.

Another issue to overcome is the balance between the efforts of providing adequate information in relation to the level of detail needed in the improvement process. In fact if a critical success factor in an improvement project driven by OEE is the retrieval of detailed information about production losses, however not all the improvement projects require a higher and really expensive data precision.

Generally there are many companies in which manual data collection is convenient. In other companies where each operator is responsible for a number of processing machines, timely and accurate data collection can be very challenging and a key goal should be fast and efficient data collection, with data put it to use throughout the day and in real-time, a more desirable approach would be realized if each machine could indicate data by itself.

An automatic OEE data recording implies:

- better accuracy;
- less labor;
- traceability;
- integrated reporting and analysis;
- immediate corrective action;
- motivation for operators.

In any case the implementation of data collection for OEE has limited value if it is not integrated in a continuous work procedure, as a part of the improvement initiative. Daily meeting and sharing information both cross-functionally and bottom-up in the organization hierarchy become a prerequisite. As well as it is useful integrating OEE into an automated management system. OEE can be applied when using a total manufacturing information system providing the detailed historical information that allows thorough diagnoses and improvement plans but more importantly it gives the summary signals.

11. Automating OEE and integration of OEE into automated management system

Automating OEE gives a company the ability to collect and classify data from the shop floor into meaningful information that can help managers understand the root causes of production inefficiency. Therefore giving greater visibility to make more informed decisions on process improvement. An automated OEE system addresses the three primary functions of OEE:

- **Acquisition:** it concerns data collection that as discussed above data will be completely automatic.
- **Analysis:** it usually provides algorithms to calculate OEE and other items related to. Moreover it is often able to support downtime classification via reason trees and other technical analysis. The more sophisticated the package, the more analysis equipment is available.
- **Visualization:** OEE metrics are available through reports or they can be displayed even via a software interface directly to the operator.

There is a lot of commercial software that provide automated OEE system, but it is possible even to integrate OEE into general tools as ERP ones. They usually offer a wide range of capabilities. They are able to gather and coordinate the operations of a plant and provide measurable information. The advantages are that database are completely integrated so the coordination among different functions involved is better. For example manufacturing can see the upcoming planned maintenance and maintenance can see the production schedules. Automated Management systems are naturally and inherently eligible for providing feasible decision support on plant profitability and establish a foundation for addressing other manufacturing challenges in the future.

12. OEE applications

At the end of the chapter, industrial examples of OEE application are presented to remark as different industries and different goals can be all involved through the OOE metric.

12.1. Case study 1

Sigma/Q [34] is a leading manufacturer of quality packaging in Northland Central America serving various markets across the globe. The company's primary goal was to improve plant performance and reduce operational costs.

The solution was to build a foundation for continuous improvement through OEE. The first step was to automate the data collection and analysis processes and introduce a real-time strategy. But the real key success factor was operator involvement in the performance improvement process. The company identified key contributors to reward them appropriately during performance reviews.

As a result, OEE increased by 40%, variability in run speed due to frequent starts and stops in the manufacturing process, was dramatically reduced and run speed was increased by 23%. Last but not least operators aspired to achieve higher levels of operational excellence, promoting a culture of continuous improvement across the various plants.

12.2. Case study 2

A global pharmaceutical company [35] has shown the will to understand if OEE as a metric could be used as an ongoing tool of improvement. It has chosen an off-shore plant and as pilot a packaging line running a full 144-hour weekly cycle and handling more than 90 products because it allowed the collection of data over both shifts. The line also had counters on most unit operations that could be easily utilized for the collection of quality data by the line operators. Twelve weeks of data was collected with operator buy-in. The test has shown that many of the current metrics were too high-level to extract the causes of issues and therefore target improvements to them. Therefore the more than 90 products routed through the test line were divided into six groups based on the highest pack rates. The continuous real-time monitoring was able to account the 90% of available run time for with little impact running the line.

12.3. Case study 3

A company providing a broad range of services to leading original equipment manufacturers in the information technology and communications industries [36] obtained three new plants from a major contract electronics manufacturer.

Each plant had distinct ways of identifying and determining downtime, as well as their own preferred techniques and practices. The goals were then:

- Find a common metric to measure productivity across plants

- Standardized downtime reporting among plants

The manufacturer's issues were complicated by the fact it makes about 30,000 different products out of 300,000 different parts, and adds an average of 2,000 new products into its manufacturing mix every month. With this number of products, frequent changeovers are necessary. It also becomes vital to have a scientific method to be able to compare all the different lines. The company was searching for a common framework in order to compare its three newest plants. The solution was the identification of factors leading to assembly line downtime. Companies utilizing this information can make comparisons across plants and assembly lines to improve effectiveness. The results were:

- OEE increase of 45%

- Identified 25% more downtime not found with previous methods

- Reduced costs

12.4. Case study 4

The Whirlpool Corporation's Findlay Division manufactures dishwashers for many brands in the world [37]. The demand for product is at an all-time high. The goal was then how to get more out of the facility and its equipment without making huge capital investments? And more specifically how can the maintenance department support the needs of manufacturing to achieve the company goals?

To make these improvements, the Division used OEE as a measure of their current equipment efficiency. As the company started tracking individual pieces of equipment's OEE ratings, it became apparent that there was room for improvement. The combination of fundamental maintenance practices such as Root Cause Failure analysis and a preventive and predictive maintenance system, along with very strong support from Division leadership, enabled the Findlay Division to get off the ground with the Total Productive Maintenance program. Again "it was the people that made this change possible" (Jim Dray, TPM Facilitator). The Division has been able to increase production by 21%, without any significant capital costs.

The OEE measure is an excellent KPI for use on both strategic and operational levels, if it is used correctly. When an organization holds people with knowledge and experience of the typical shortages of OEE and its common implementation challenges, the probability of achieving the intended benefits of OEE will certainly increase. Based on using OEE as an improvement driver at the case study company, some success factors have been identified:

- A standard definition of OEE must be clearly defined and communicated at all levels within the organization since this is the foundation for its utilization. It is especially important to determine how the ideal cycle time and planned and unplanned downtime should be interpreted.

- Involving the operators in the process of defining production loss causes and configuring the templates and lists to be used for monitoring promotes operator commitment, understanding of the procedure and awareness of the frequency of sporadic and chronic disturbances.

- Driving the OEE implementation as a project with a predefined organization, a structured working procedure promoting cross-functional and shop floor involvement, and practical guidance on what activities to execute and in what order, implies resource allocation that forces management attention and puts OEE on the agenda.

- Viewing and communicating OEE as a driver for improvements rather than a management measure for follow-up and control of performance (although this is also the case) is one of the cornerstones for a successful OEE implementation.

- Active involvement of the support functions, especially production engineering and maintenance, is required, otherwise the level of improvements to increase OEE will not be enough and the speed of change will consequently be too low.

- Separating improvement actions into those directly having an impact on process stability, i.e. OEE, from those with indirect impact is necessary especially in the initial implementation phase to show quick results.

- Including reporting OEE and prioritized daily actions in the routines of daily follow-up meetings (from team level to department/site level) is an excellent way to integrate OEE as a driver for improvements in the operations management system.

- Results should be communicated, e.g. by graphical visualization of the OEE improvements on the boards. Visualizing OEE and process output together are illustrative and motivating.

- Including production performance in the company´s overall production strategy and managing this with a continuous follow up of OEE as a KPI on different consolidation levels is the optimal driver for efficient management. When top management attention is continuously given to the process of achieving stable production processes the possibilities of reaching good results certainly increases.

13. Conclusion

There are many challenges associated with the implementation of OEE for monitoring and managing production performance, for example:

- how it is defined, interpreted and compared
- how the OEE data are collected and analyzed

- how it is monitored and by whom
- how it aligns with the overall production strategy
- how it could be utilized for sustainability purpose.

Moreover it is remarkable that setting high OEE goals in an environment with excessive capacity is of less value since it is not possible to utilize the equipment full time. OEE measure is less suitable as a target KPI, since OEE only measures the efficiency during the time the equipment is planned to be operating, while equipment and personnel drives manufacturing costs both when they are in operation and during downtime.

The purpose of measuring OEE can be questioned in the light of the financial crisis. There are some authors that have reported the need of further research work on linking OEE with financial measures. Dal et al. [31] asserts "there would appear to be a useful line of research in exploring the link between OEE and the popular business models such as balanced scorecard". Muchiri et al. [16] suggests "Further research should explore the dynamics of translating equipment effectiveness or loss of effectiveness in terms of cost." The authors agree with these statements, there is clearly a missing link between OEE and manufacturing cost. Jonsson et al. [39] presents a manufacturing cost model linking production performance with economic parameters. The utilization of this manufacturing cost model in developing industrially applicable productivity KPI´s will be elaborated on in future research.

Author details

Raffaele Iannone[1] and Maria Elena Nenni[2*]

*Address all correspondence to: menenni@unina.it

1 Department of Industrial Engineering, University of Salerno, Italy

2 Department of Industrial Engineering, University of Naples Federico II, Italy

References

[1] Fleischer, J, Weismann, U, & Niggeschmidt, S. Calculation and Optimisation Model for Costs and Effects of Availability Relevant Service Elements: proceedings of the CIRP International Conference on Life Cycle Engineering, LCE2006, 31 May- 2 June (2006). Leuven, Belgium.

[2] Huang, S. H, Dismukes, J. P, Mousalam, A, Razzak, R. B, & Robinson, D. E. Manufacturing Productivity improvement using effectiveness metrics and simulation analysis. International Journal of Production Research (2003). , 41(3), 513-527.

[3] Nakajima, S. Introduction to TPM: Total Productive Maintenance. Productivity Press; (1988).

[4] Womack, J. P, Jones, D. T, & Roos, D. The Machine That Changed the World. Rawson Associates; (1990).

[5] Harry, M. J. Six Sigma: a breakthrough strategy for profitability. Quality Progress (1998).

[6] Todd, J. World-class Manufacturing. McGraw-Hill; (1995).

[7] Nakajima, S. TPM Development Program, Productivity Press; (1989).

[8] E79-98 Guideline for the Definition and Measurements of Overall Equipment Effectiveness: proceedings of Advanced Semiconductor Manufacturing Conference and Workshop, IEEE/SEMI 1998, 23- 25 September 1998, Boston, MA.

[9] Koch, A. OEE for Operators: Overall Equipment Effectiveness. Productivity Press; (1999).

[10] Hansen, B. Overall Equipment Effectiveness. Industrial Press; (2001).

[11] Stamatis, D. H. The OEE Primer. Understanding Overall Equipment Effectiveness, Reliability, and Maintainability. Taylor & Francis; (2010).

[12] Jonsson, P, & Lesshammar, M. Evaluation and improvement of manufacturing performance measurement systems- the role of OEE. International Journal of Operations & Production Management (1999). , 19(1), 55-78.

[13] Louglin, S. Aholistic Approach to Overall Equipment Effectiveness. IEE Computing and Control Engineering Journal (2003). , 14(6), 37-42.

[14] Bicheno, J. The New Lean Toolbox towards fast flexible flow. Moreton Press; (2004).

[15] Andersson, C, & Bellgran, M. Managing Production Performance with Overall Equipment Efficiency (OEE)- Implementation Issues and Common Pitfalls. http://msep.engr.wisc.edu/phocadownload/cirp44_managing%20production%20performance.pdfaccessed 20 September (2012).

[16] Muchiri, P, & Pintelon, L. Performance measurement using overall equipment effectiveness (OEE): Literature review and practical application discussion. International Journal of Production Research (2008). , 46(13), 1-45.

[17] Ivancic, I. Development of Maintenance in Modern Production: proceedings of 14th European Maintenance Conference, EUROMAINTENANCE'October (1998). Dubrovnik, Hrvatska., 98, 5-7.

[18] Raouf, A. Improving Capital Productivity Through Maintenance. International Journal of Operations & Production Management (1994). , 14(7), 44-52.

[19] Oechsner, R. From OEE to OFE. Materials science in semiconductor processing (2003).

[20] Huang, S. H, & Keskar, H. Comprehensive and configurable metrics for supplier selection. International Journal of Production Economics (2007).

[21] Scott, D, & Pisa, R. Can Overall Factory Effectiveness Prolong Moore's Law? Solid State Technology (1998).

[22] Goldratt, E. M, & Cox, J. The Goal: A Process of Ongoing Improvement. North River Press; (1992).

[23] Huang, S. H, Dimukes, J. P, Shi, J, Su, Q, Razzak, M. A, & Robinson, D. E. Manufacturing system modeling for productivity improvement. Journal of Manufacturing Systems (2002). , 2002, 21-249.

[24] Muthiah, K. M. N, Huang, S. H, & Mahadevan, S. Automating factory performance diagnostics using overall throughput effectiveness (OTE) metric. International Journal of Advanced Manufacturing Technology (2008).

[25] Bose, S. J. An Introduction to Queueing Systems. Kluwer/Plenum Publishers; (2002).

[26] Meyn, S. P, & Tweedie, R. L. Markov Chains and Stochastic Stability. Springer-Verlag; (1993).

[27] David, R, & Alla, H. Discrete, continuous, and hybrid Petri Nets. Springer; (2005).

[28] Cheng-leong, A, Pheng, K. L, & Leng, G. R. K. IDEF: a comprehensive modelling methodology for the development of manufacturing enterprise systems. International Journal of Production Research (1999). , 1999, 37-3839.

[29] Santarek, K, & Buseif, I. M. Modeling and design of flexible manufacturing systems using SADT and petri nets tools. Journal of Material Process Technology (1998). , 1998, 76-212.

[30] Muthiah, K. M. N, & Huang, S. H. A review of literature on manufacturing systems productivity measurement and improvement. International Journal of Industrial Engineering (2006). , 2006, 1-461.

[31] Dal, B, Tugwell, P, & Greatbanks, R. Overall equipment effectiveness as a measure of operational improvement. A practical analysis. International Journal of Operations & Production Management (2000).

[32] Deming, W. E. Out of the Crisis. MIT Center for Advanced Engineering Study; (1986).

[33] Ericsson, J. Disruption analysis- An Important Tool in Lean Production. PhD thesis. Department of Production and Materials Engineering, Lund University, Sweden; (1997).

[34] http://wwwneustro.com/oeedocs/Sigma_Q_Neustro.pdf (accessed 3 October (2012).

[35] http://wwwinformance.com/download.aspx?id=pharma_casestudy.pdf (accessed 12 October (2012).

[36] http://wwwinformance.com/download.aspx?id=CASE-STUDY_HIGH-TECH.pdf (accessed 20 October (2012).

[37] http://wwwleanexpertise.com/TPMONLINE/articles_on_total_productive_maintenance/tpm/whirpoolcase.htm (accessed 12 October (2012).

[38] Jonsson, M, Andersson, C, & Stahl, J. E. A general economic model for manufacturing cost simulation: proceedings of the 41st CIRP Conference on Manufacturing Systems, May (2008). Tokyo., 26-28.

Chapter 4

Reliability and Maintainability in Operations Management

Filippo De Carlo

Additional information is available at the end of the chapter

1. Introduction

The study of component and process reliability is the basis of many efficiency evaluations in Operations Management discipline. For example, in the calculation of the Overall Equipment Effectiveness (OEE) introduced by Nakajima [1], it is necessary to estimate a crucial parameter called availability. This is strictly related to reliability. Still as an example, consider how, in the study of service level, it is important to know the availability of machines, which again depends on their reliability and maintainability.

Reliability is defined as the probability that a component (or an entire system) will perform its function for a specified period of time, when operating in its design environment. The elements necessary for the definition of reliability are, therefore, an unambiguous criterion for judging whether something is working or not and the exact definition of environmental conditions and usage. Then, reliability can be defined as the time dependent probability of correct operation if we assume that a component is used for its intended function in its design environment and if we clearly define what we mean with "failure". For this definition, any discussion on the reliability basics starts with the coverage of the key concepts of probability.

A broader definition of reliability is that "reliability is the science to predict, analyze, prevent and mitigate failures over time." It is a science, with its theoretical basis and principles. It also has sub-disciplines, all related - in some way - to the study and knowledge of faults. Reliability is closely related to mathematics, and especially to statistics, physics, chemistry, mechanics and electronics. In the end, given that the human element is almost always part of the systems, it often has to do with psychology and psychiatry.

In addition to the prediction of system durability, reliability also tries to give answers to other questions. Indeed, we can try to derive from reliability also the availability performance of a system. In fact, availability depends on the time between two consecutive failures and on how long it takes to restore the system. Reliability study can be also used to understand how faults can be avoided. You can try to prevent potential failures, acting on the design, materials and maintenance.

Reliability involves almost all aspects related to the possession of a property: cost management, customer satisfaction, the proper management of resources, passing through the ability to sell products or services, safety and quality of the product.

This chapter presents a discussion of reliability theory, supported by practical examples of interest in operations management. Basic elements of probability theory, as the sample space, random events and Bayes' theorem should be revised for a deeper understanding.

2. Reliability basics

The period of regular operation of an equipment ends when any chemical-physical phenomenon, said fault, occurred in one or more of its parts, determines a variation of its nominal performances. This makes the behavior of the device unacceptable. The equipment passes from the state of operation to that of non-functioning.

In Table 1 faults are classified according to their origin. For each failure mode an extended description is given.

Failure cause	Description
Stress, shock, fatigue	Function of the temporal and spatial distribution of the load conditions and of the response of the material. The structural characteristics of the component play an important role, and should be assessed in the broadest form as possible, incorporating also possible design errors, embodiments, material defects, etc..
Temperature	Operational variable that depends mainly on the specific characteristics of the material (thermal inertia), as well as the spatial and temporal distribution of heat sources.
Wear	State of physical degradation of the component; it manifests itself as a result of aging phenomena that accompany the normal activities (friction between the materials, exposure to harmful agents, etc..)
Corrosion	Phenomenon that depends on the characteristics of the environment in which the component is operating. These conditions can lead to material degradation or chemical and physical processes that make the component no longer suitable.

Table 1. Main causes of failure. The table shows the main cases of failure with a detailed description

To study reliability you need to transform reality into a model, which allows the analysis by applying laws and analyzing its behavior [2]. Reliability models can be divided into static and dynamic ones. **Static models** assume that a failure does not result in the occurrence of other faults. **Dynamic reliability**, instead, assumes that some failures, so-called primary failures, promote the emergence of secondary and tertiary faults, with a cascading effect. In this text we will only deal with static models of reliability.

In the traditional paradigm of static reliability, individual components have a binary status: either working or failed. Systems, in turn, are composed by an integer number n of components, all mutually independent. Depending on how the components are configured in creating the system and according to the operation or failure of individual components, the system either works or does not work.

Let's consider a generic X system consisting of n elements. The static reliability modeling implies that the operating status of the i - th component is represented by the state function X_i defined as:

$$X_i = \begin{cases} 1 & \text{if the } i \text{ - th component works} \\ 0 & \text{if the } i \text{ - th component fails} \end{cases}$$

(1)

The state of operation of the system is modeled by the state function $\Phi(X)$

$$\Phi(X) = \begin{cases} 1 & \text{if the system works} \\ 0 & \text{if the system fails} \end{cases}$$

(2)

The most common configuration of the components is the series system. A series system works if and only if all components work. Therefore, the status of a series system is given by the state function:

$$\Phi(X) = \prod_{i=1}^{n} X_i = \min_{i \in \{1,2...,n\}} X_i$$

(3)

where the symbol $\prod$ indicates the product of the arguments.

System configurations are often represented graphically with Reliability Block Diagrams (RBDs) where each component is represented by a block and the connections between them express the configuration of the system. The operation of the system depends on the ability to cross the diagram from left to right only by passing through the elements in operation. Figure 1 contains the RBD of a four components series system.

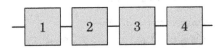

Figure 1. Reliability block diagram for a four components (1,2,3,4) series system.

The second most common configuration of the components is the parallel system. A parallel system works if and only if at least one component is working. A parallel system does not work if and only if all components do not work. So, if $\bar{\Phi}(X)$ is the function that represents the state of not functioning of the system and $\bar{X}_i$ indicates the non-functioning of the i - th element, you can write:

$$\bar{\Phi}(X)=\prod_{i=1}^{n} \bar{X}_i \tag{4}$$

Accordingly, the state of a parallel system is given by the state function:

$$\Phi(X)=1-\prod_{i=1}^{n}\left(1-X_i\right)=\coprod_{i=1}^{n} X_i = \max_{i\in\{1,2,\dots,n\}} X_i \tag{5}$$

where the symbol $\coprod$ indicates the complement of the product of the complements of the arguments. Figure 2 contains a RBD for a system of four components arranged in parallel.

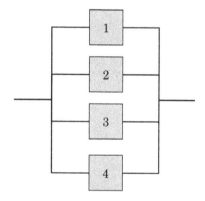

Figure 2. Parallel system. The image represents the RBD of a system of four elements (1,2,3,4) arranged in a reliability parallel configuration.

Another common configuration of the components is the series-parallel systems. In these systems, components are configured using combinations in series and parallel configurations. An example of such a system is shown in Figure 3.

State functions for series-parallel systems are obtained by decomposition of the system. With this approach, the system is broken down into subsystems or configurations that are in series or in parallel. The state functions of the subsystems are then combined appropriately, depending on how they are configured. A schematic example is shown in Figure 4.

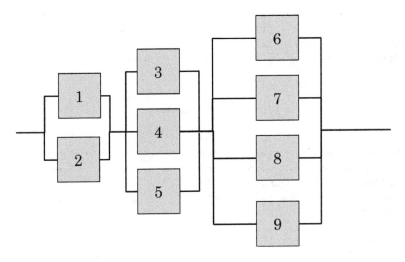

Figure 3. Series-parallel system. The picture shows the RBD of a system due to the series-parallel model of 9 elementary units.

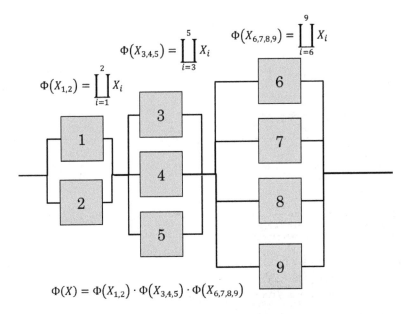

Figure 4. Calculation of the state function of a series-parallel. Referring to the configuration of Figure 3, the state function of the system is calculated by first making the state functions of the parallel of{1,2}, of {3,4, 5} and of {6,7, 8 , 9}. Then we evaluate the state function of the series of the three groups just obtained.

A particular component configuration, widely recognized and used, is the **parallel k out of** n. A system k out of n works if and only if at least k of the n components works. Note that a series system can be seen as a system n out of n and a parallel system is a system 1 out of n. The state function of a system k out of n is given by the following algebraic system:

$$\Phi(X) = \begin{cases} 1 & \text{if } \sum_{i=1}^{n} X_i \geq k \\ 0 & \text{otherwise} \end{cases} \tag{6}$$

The RBD for a system k out of n has an appearance identical to the RBD schema of a parallel system of n components with the addition of a label "k out of n". For other more complex system configurations, such as the bridge configuration (see Figure 5), we may use more intricate techniques such as the minimal path set and the minimal cut set, to construct the system state function.

A Minimal Path Set - MPS is a subset of the components of the system such that the operation of all the components in the subset implies the operation of the system. The set is minimal because the removal of any element from the subset eliminates this property. An example is shown in Figure 5.

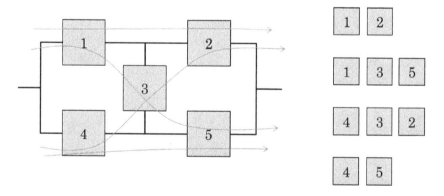

Figure 5. Minimal Path Set. The system on the left contains the minimal path set indicated by the arrows and shown in the right part. Each of them represents a minimal subset of the components of the system such that the operation of all the components in the subset implies the operation of the system.

A Minimal Cut Set - MCS is a subset of the components of the system such that the failure of all components in the subset does not imply the operation of the system. Still, the set is called minimal because the removal of any component from the subset clears this property (see Figure 6).

MCS and MPS can be used to build equivalent configurations of more complex systems, not referable to the simple series-parallel model. The first equivalent configuration is based on the consideration that the operation of all the components, in at least a MPS, entails the operation

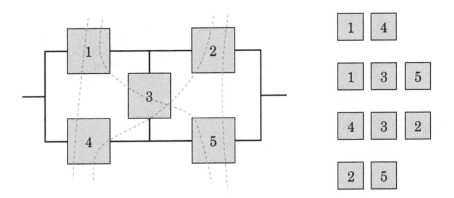

Figure 6. Minimal Cut Set. The system of the left contains the minimal cut set, indicated by the dashed lines, shown in the right part. Each of them represents a minimum subset of the components of the system such that the failure of all components in the subset does not imply the operation of the system.

of the system. This configuration is, therefore, constructed with the creation of a series subsystem for each path using only the minimum components of that set. Then, these subsystems are connected in parallel. An example of an equivalent system is shown in Figure 7.

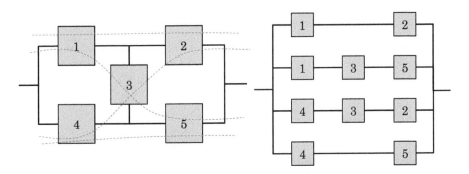

Figure 7. Equivalent configurations with MPS. You build a series subsystem for each MPS. Then such subsystems are connected in parallel.

The second equivalent configuration, is based on the logical principle that the failure of all the components of any MCS implies the fault of the system. This configuration is built with the creation of a parallel subsystem for each MCS using only the components of that group. Then, these subsystems are connected in series (see Figure 8).

After examining the components and the status of the system, the next step in the static modeling of reliability is that of considering the probability of operation of the component and of the system.

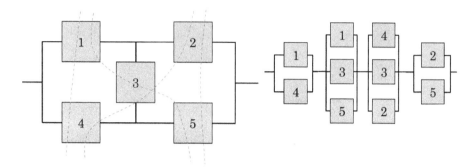

Figure 8. Equivalent configurations with MCS. You build a subsystem in parallel for each MCS. Then the subsystems are connected in series.

The reliability R_i of the i - th component is defined by:

$$R_i = P(X_i = 1) \qquad (7)$$

while the **reliability of the system** R is defined as in equation 8:

$$R = P(\Phi(X) = 1) \qquad (8)$$

The methodology used to calculate the reliability of the system depends on the configuration of the system itself. For a series system, the reliability of the system is given by the product of the individual reliability (law of Lusser, defined by German engineer Robert Lusser in the 50s):

$$R = \prod_{i=1}^{n} R_i \quad \text{since} \quad R = P\left(\bigcap_{i=1}^{n}(X_i = 1)\right) = \prod_{i=1}^{n} P(X_i = 1) = \prod_{i=1}^{n} R_i \qquad (9)$$

For an example, see Figure 9.

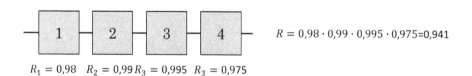

$$R = 0,98 \cdot 0,99 \cdot 0,995 \cdot 0,975 = 0,941$$

$R_1 = 0,98 \quad R_2 = 0,99 \; R_3 = 0,995 \quad R_3 = 0,975$

Figure 9. serial system consisting of 4 elements with reliability equal to 0.98, 0.99, 0.995 and 0.975. The reliability of the whole system is given by their product: $R = 0.98 \cdot 0.99 \cdot 0.995 \cdot 0.975 = 0.941$

For a parallel system, reliability is:

$$R = 1 - \prod_{i=1}^{n}(1 - R_i) = \coprod_{i=1}^{n} R_i \qquad (10)$$

In fact, from the definition of system reliability and by the properties of event probabilities, it follows:

$$R = P\left(\bigcup_{i=1}^{n}(X_i = 1)\right) = 1 - P\left(\bigcap_{i=1}^{n}(X_i = 0)\right) = 1 - \prod_{i=1}^{n}P(X_i = 0) = = 1 - \prod_{i=1}^{n}[1 - P(X_i = 1)] = 1 - \prod_{i=1}^{n}(1 - R_i) = \coprod_{i=1}^{n} R_i \qquad (11)$$

In many parallel systems, components are identical. In this case, the reliability of a parallel system with n elements is given by:

$$R = 1 - (1 - R_i)^n \qquad (12)$$

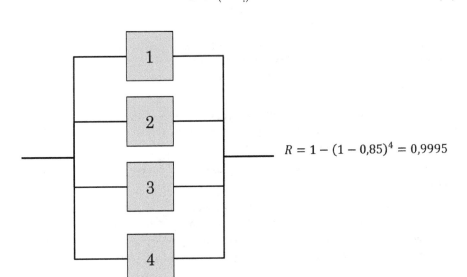

$$R = 1 - (1 - 0{,}85)^4 = 0{,}9995$$

$$R_1 = R_2 = R_3 = R_4 = 0{,}85$$

Figure 10. A parallel system consisting of 4 elements with the same reliability of 0.85. The system reliability s given by their co-product: $1 - (1 - 0.85)^4 = 0.9995$.

For a series-parallel system, system reliability is determined using the same approach of decomposition used to construct the state function for such systems. Consider, for instance, the system drawn in Figure 11, consisting of 9 elements with reliability $R_1 = R_2 = 0.9$; $R_3 = R_4 = R_5 = 0.8$ and $R_6 = R_7 = R_8 = R_9 = 0.7$. Let's calculate the overall reliability of the system.

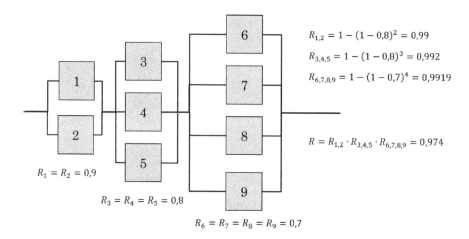

$$R_{1,2} = 1 - (1 - 0{,}8)^2 = 0{,}99$$

$$R_{3,4,5} = 1 - (1 - 0{,}8)^3 = 0{,}992$$

$$R_{6,7,8,9} = 1 - (1 - 0{,}7)^4 = 0{,}9919$$

$$R = R_{1,2} \cdot R_{3,4,5} \cdot R_{6,7,8,9} = 0{,}974$$

$$R_1 = R_2 = 0{,}9$$

$$R_3 = R_4 = R_5 = 0{,}8$$

$$R_6 = R_7 = R_8 = R_9 = 0{,}7$$

Figure 11. The system consists of three groups of blocks arranged in series. Each block is, in turn, formed by elements in parallel. First we must calculate $R_{1,2}=1-(1-0.8)^2=0.99$. So it is possible to estimated $R_{3,4,5}=1-(1-0.8)^3=0.992$. Then we must calculate the reliability of the last parallel block $R_{6,7,8,9}=1-(1-0.7)^4=0.9919$. Finally, we proceed to the series of the three blocks: $R=R_{1,2} \cdot R_{3,4,5} \cdot R_{6,7,8,9}=0.974$.

To calculate the overall reliability, for all other types of systems which can't be brought back to a series-parallel scheme, it must be adopted a more intensive calculation approach [3] that is normally done with the aid of special software.

Reliability functions of the system can also be used to calculate measures of **reliability importance**.

These measurements are used to assess which components of a system offer the greatest opportunity to improve the overall reliability. The most widely recognized definition of reliability importance I'_i of the components is the **reliability marginal gain**, in terms of overall system rise of functionality, obtained by a marginal increase of the component reliability:

$$I'_i = \frac{\partial R}{\partial R_i} \tag{13}$$

For other system configurations, an alternative approach facilitates the calculation of reliability importance of the components. Let $R(1_i)$ be the reliability of the system modified so that $R_i=1$ and $R(0_i)$ be the reliability of the system modified with $R_i=0$, always keeping unchanged the other components. In this context, the reliability importance I_i is given by:

$$I_i = R(1_i) - R(0_i) \tag{14}$$

In a series system, this formulation is equivalent to writing:

$$I_i = \prod_{\substack{j=1 \\ j \neq i}}^{n} R_j \qquad (15)$$

Thus, the most important component (in terms of reliability) in a series system is the less reliable. For example, consider three elements of reliability $R_1 = 0.9$, $R_2 = 0.8$ e $R_3 = 0.7$. It is therefore: $I_1 = 0.8 \bullet 0.7 = 0.56$, $I_2 = 0.9 \bullet 0.7 = 0.63$ and $I_3 = 0.9 \cdot 0.8 = 0.72$ which is the higher value.

If the system is arranged in parallel, the reliability importance becomes as follows:

$$I_i = \prod_{\substack{j=1 \\ j \neq i}}^{n} \left(1 - R_j\right) \qquad (16)$$

It follows that the most important component in a parallel system is the more reliable. With the same data as the previous example, this time having a parallel arrangement, we can verify Eq. 16 for the first item: $I_1 = R(1_1) - R(0_1) = [1 - (1 - 1) \cdot (1 - 0.8) \bullet (1 - 0.7)] - [1 - (1 - 0) \cdot (1 - 0.8) \bullet (1 - 0.7)]$ $= 1 - 0 - 1 + (1 - 0.8) \bullet (1 - 0.7) = (1 - 0.8) \bullet (1 - 0.7)$.

For the calculation of the reliability importance of components belonging to complex systems, which are not attributable to the series-parallel simple scheme, reliability of different systems must be counted. For this reason the calculation is often done using automated algorithms.

3. Fleet reliability

Suppose you have studied the reliability of a component, and found that it is 80% for a mission duration of 3 hours. Knowing that we have 5 identical items simultaneously active, we might be interested in knowing what the overall reliability of the group would be. In other words, we want to know what is the probability of having a certain number of items functioning at the end of the 3 hours of mission. This issue is best known as fleet reliability.

Consider a set of m identical and independent systems in a same instant, each having a reliability R. The group may represent a set of systems in use, independent and identical, or could represent a set of devices under test, independent and identical. A discrete random variable of great interest reliability is N, the number of functioning items. Under the assumptions specified, N is a binomial random variable, which expresses the probability of a Bernoulli process. The corresponding probabilistic model is, therefore, the one that describes the extraction of balls from an urn filled with a known number of red and green balls. Suppose that the percentage R of green balls is coincident with the reliability after 3 hours. After each extraction from the urn, the ball is put back in the container. Extraction is repeated m times, and we look for the probability of finding n green. The sequence of random variables thus obtained is a Bernoulli process of which each extraction is a test. Since the probability of obtaining N successes in m extractions from an urn, with restitution of the ball, follows the binomial distribution $B(m, R)B$, the probability mass function of N is the well-known:

$$P(N=n)=\frac{m!}{n!(m-n)!}R^{n}(1-R)^{m-n} \tag{17}$$

The expected value of N is given by: $E(N)=\mu_N=m \bullet R$ and the standard deviation is: $\sigma_N=\sqrt{m \bullet R \bullet (1-R)}$.

Let's consider, for example, a corporate fleet consisting of 100 independent and identical systems. All systems have the same mission, independent from the other missions. Each system has a reliability of mission equal to 90%. We want to calculate the average number of missions completed and also what is the probability that at least 95% of systems would complete their mission. This involves analyzing the distribution of the binomial random variable characterized by $R = 0.90$ and $m = 100$. The expected value is given by $E(N)=\mu_N=100 \bullet 0.9=90$.

The probability that at least 95% of the systems complete their mission can be calculated as the sum of the probabilities that complete their mission 95, 96, 97, 98, 99 and 100 elements of the fleet:

$$P(N \geq n)=\sum_{n=95}^{100}\left[\frac{m!}{n!(m-n)!}R^{n}(1-R)^{m-n}\right]=0,058 \tag{18}$$

4. Time dependent reliability models

When reliability is expressed as a function of time, the continuous random variable, not negative, of interest is T, the instant of failure of the device. Let $f(t)$ be the probability density function of T, and let $F(t)$ be the cumulative distribution function of T. $F(t)$ is also known as failure function or unreliability function [4].

In the context of reliability, two additional functions are often used: the **reliability** and the hazard function. Let's define **Reliability** $R(t)$ as the survival function:

$$R(t)=P(T \geq t)=1-F(t) \tag{19}$$

The **Mean Time To Failure - MTTF** is defined as the expected value of the failure time:

$$MTTF = E(T)=\int_0^{\infty}t \bullet f(t) \bullet dt \tag{20}$$

Integrating by parts, we can prove the equivalent expression:

$$MTTF = E(T)=\int_0^{\infty}R(t) \bullet dt \tag{21}$$

5. Hazard function

Another very important function is the **hazard function**, denoted by $\lambda(t)$, defined as the trend of the instantaneous failure rate at time t of an element that has survived up to that time t. The failure rate is the ratio between the instantaneous probability of failure in a neighborhood of t- conditioned to the fact that the element is healthy in t- and the amplitude of the same neighborhood.

The hazard function $\lambda(t)$ [5] coincides with the intensity function $z(t)$ of a Poisson process. The hazard function is given by:

$$\lambda(t) = \lim_{\Delta t \to 0} \frac{P(t \leq T < t + \Delta t \mid T \geq t)}{\Delta t} \tag{22}$$

Thanks to Bayes' theorem, it can be shown that the relationship between the hazard function, density of probability of failure and reliability is the following:

$$\lambda(t) = \frac{f(t)}{R(t)} \tag{23}$$

Thanks to the previous equation, with some simple mathematical manipulations, we obtain the following relation:

$$R(t) = e^{-\int_0^t \lambda(u) \bullet du} \tag{24}$$

In fact, since $\ln[R(0)] = \ln[1] = 0$, we have:

$$R(t) = \frac{f(t)}{\lambda(t)} = \frac{1}{\lambda(t)} \bullet \frac{dF(t)}{dt} = -\frac{1}{\lambda(t)} \bullet \frac{dR(t)}{dt} \to \frac{1}{R(t)} dR(t) = -\lambda(t) dt \to \ln[R(t)] - \ln[R(0)] = -\int_0^t \lambda(u) du \tag{25}$$

From equation 24 derive the other two fundamental relations:

$$F(t) = 1 - e^{-\int_0^t \lambda(u) \bullet du} \qquad f(t) = \lambda(t) \bullet e^{-\int_0^t \lambda(u) \bullet du} \tag{26}$$

The most popular conceptual model of the hazard function is the **bathtub curve**. According to this model, the failure rate of the device is relatively high and descending in the first part of the device life, due to the potential manufacturing defects, called **early failures**. They manifest themselves in the first phase of operation of the system and their causes are often linked to structural deficiencies, design or installation defects. In terms of reliability, a system that manifests infantile failures improves over the course of time.

Later, at the end of the life of the device, the failure rate increases due to wear phenomena. They are caused by alterations of the component for material and structural aging. The beginning of the period of wear is identified by an increase in the frequency of failures which continues as time goes by. The **wear-out failures** occur around the average age of operating; the only way to avoid this type of failure is to replace the population in advance.

Between the period of early failures and of wear-out, the failure rate is about constant: failures are due to random events and are called **random failures**. They occur in non-nominal operating conditions, which put a strain on the components, resulting in the inevitable changes and the consequent loss of operational capabilities. This type of failure occurs during the useful life of the system and corresponds to unpredictable situations. The central period with constant failure rate is called **useful life**. The juxtaposition of the three periods in a graph which represents the trend of the failure rate of the system, gives rise to a curve whose characteristic shape recalls the section of a bathtub, as shown in Figure 12.

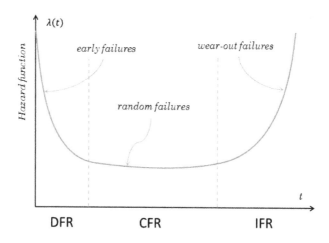

Figure 12. Bathtub curve. The hazard function shape allows us to identify three areas: the initial period of the early failures, the middle time of the useful life and the final area of wear-out.

The most common mathematical classifications of the hazard curve are the so called **Constant Failure Rate - CFR, Increasing Failure Rate - IFR** and **Decreasing Failure Rate - DFR**.

The CFR model is based on the assumption that the failure rate does not change over time. Mathematically, this model is the most simple and is based on the principle that the faults are purely random events. The IFR model is based on the assumption that the failure rate grows up over time. The model assumes that faults become more likely over time because of wear, as is frequently found in mechanical components. The DFR model is based on the assumption that the failure rate decreases over time. This model assumes that failures become less likely as time goes by, as it occurs in some electronic components.

Since the failure rate may change over time, one can define a reliability parameter that behaves as if there was a kind of counter that accumulates hours of operation. The **residual reliability** function $R(t + t_0 \mid t_0)$, in fact, measures the reliability of a given device which has already survived a determined time t_0. The function is defined as follows:

$$R(t + t_0 \mid t_0) = P(T > t + t_0 \mid T > t_0) \tag{27}$$

Applying Bayes' theorem we have:

$$P(T > t + t_0 \mid T > t_0) = \frac{P(T > t_0 \mid T > t + t_0) \bullet P(T > t + t_0)}{P(T > t_0)} \tag{28}$$

And, given that $P(T > t_0 \mid T > t + t_0) = 1$, we obtain the final expression, which determines the residual reliability:

$$R(t + t_0 \mid t_0) = \frac{R(t + t_0)}{R(t_0)} \tag{29}$$

The **residual Mean Time To Failure – residual MTTF** measures the expected value of the residual life of a device that has already survived a time t_0:

$$MTTF(t_0) = E(T - t_0 \mid T > t_0) = \int_0^\infty R(t + t_0 \mid t_0) \bullet dt \tag{30}$$

For an IFR device, the residual reliability and the residual MTTF, decrease progressively as the device accumulates hours of operation. This behavior explains the use of preventive actions to avoid failures. For a DFR device, both the residual reliability and the residual MTTF increase while the device accumulates hours of operation. This behavior motivates the use of an intense running (burn-in) to avoid errors in the field.

The **Mean Time To Failure –MTTF**, measures the expected value of the life of a device and coincides with the residual time to failure, where $t_0 = 0$. In this case we have the following relationship:

$$MTTF = MTTF(0) = E(T \mid T > 0) = \int_0^\infty R(t) \bullet dt \tag{31}$$

The **characteristic life** of a device is the time t_C corresponding to a reliability $R(t_C)$ equal to $1/e$, that is the time for which the area under the hazard function is unitary:

$$R(t_C) = e^{-1} = 0,368 \rightarrow R(t_C) = \int_0^{t_C} \lambda(u) \bullet du = 1 \tag{32}$$

Let us consider a CFR device with a constant failure rate λ. The time-to-failure is an exponential random variable. In fact, the probability density function of a failure, is typical of an exponential distribution:

$$f(t) = \lambda(t) \bullet e^{-\int_0^t \lambda(u) \bullet du} = \lambda e^{-\lambda \bullet t} \tag{33}$$

The corresponding cumulative distribution function $F(t)$ is:

$$F(t) = \int_{-\infty}^t f(z)dz = \int_{-\infty}^t \lambda e^{-\lambda \bullet z} dz = 1 - e^{-\lambda \bullet t} \tag{34}$$

The reliability function $R(t)$ is the survival function:

$$R(t) = 1 - F(t) = e^{-\lambda \bullet t} \tag{35}$$

For CFR items, the residual reliability and the residual MTTF both remain constant when the device accumulates hours of operation. In fact, from the definition of residual reliability, $\forall\, t_0 \in [0, \infty]$, we have:

$$R(t + t_0 \mid t_0) = \frac{R(t + t_0)}{R(t_0)} = \frac{e^{-\lambda \bullet (t + t_0)}}{e^{-\lambda \bullet t_0}} = e^{-\lambda \bullet (t + t_0) + \lambda \bullet t_0} = e^{-\lambda \bullet t} = R(t) \tag{36}$$

Similarly, for the residual MTTF, is true the invariance in time:

$$MTTF(t_0) = \int_0^\infty R(t + t_0 \mid t_0) \bullet dt = \int_0^\infty R(t) \bullet dt \qquad \forall\, t_0 \in [0, \infty] \tag{37}$$

This behavior implies that the actions of prevention and running are useless for CFR devices. Figure 13 shows the trend of the function $f(t) = \lambda \bullet e^{-\lambda \bullet t}$ and of the cumulative distribution function $F(t) = 1 - e^{-\lambda \bullet t}$ for a constant failure rate $\lambda = 1$. In this case, since $\lambda = 1$, the probability density function and the reliability function, overlap: $f(t) = R(t) = e^{-t}$.

The probability of having a fault, not yet occurred at time t, in the next dt, can be written as follows:

$$P(t < T < t + dt \mid T > t) \tag{38}$$

Recalling the Bayes' theorem, in which we consider the probability of an hypothesis H, being known the evidence E:

$$P(H \mid E) = \frac{P(E \mid H) \bullet P(H)}{P(E)} \tag{39}$$

we can replace the evidence E with the fact that the fault has not yet taken place, from which we obtain $P(E) \rightarrow P(T > t)$. We also exchange the hypothesis H with the occurrence of the fault in the neighborhood of t, obtaining $P(H) \rightarrow P(t < T < t + dt)$. So we get:

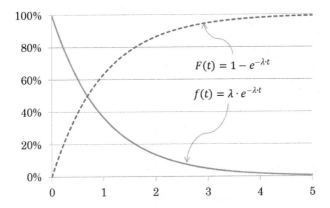

Figure 13. Probability density function and cumulative distribution of an exponential function. In the figure is seen the trend of $f(t)=\lambda \bullet e^{-\lambda \bullet t}$ and of $f(t)=\lambda \bullet e^{-\lambda \bullet t}$ with $\lambda = 1$.

$$P(t<T<t+dt \mid T>t)=\frac{P(T>t \mid t<T<t+dt) \bullet P(t<T<t+dt)}{P(T>t)} \tag{40}$$

Since $P(T>t \mid t<T<t+dt)=1$, being a certainty, it follows:

$$P(t<T<t+dt \mid T>t)=\frac{P(t<T<t+dt)}{P(T>t)}=\frac{f(t)dt}{e^{-\lambda \bullet t}}=\frac{\lambda e^{-\lambda \bullet t}dt}{e^{-\lambda \bullet t}}=\lambda \bullet dt \tag{41}$$

As can be seen, this probability does not depend on t, i.e. it is not function of the life time already elapsed. It is as if the component does not have a memory of its own history and it is for this reason that the exponential distribution is called **memoryless**.

The use of the constant failure rate model, facilitates the calculation of the characteristic life of a device. In fact for a CFR item, t_C is the reciprocal of the failure rate. In fact:

$$R(t_C)=e^{-\lambda \bullet t_C}=e^{-1} \rightarrow t_C=\frac{1}{\lambda} \tag{42}$$

Therefore, the characteristic life, in addition to be calculated as the time value t_C for which the reliability is 0.368, can more easily be evaluated as the reciprocal of the failure rate.

The definition of MTTF, in the CFR model, can be integrated by parts and give:

$$MTTF=\int_0^\infty R(t) \bullet dt=\int_0^\infty e^{-\lambda \bullet t} \bullet dt=-\frac{1}{\lambda}e^{-\lambda \bullet t}\Big|_0^\infty = -\frac{0}{\lambda}+\frac{1}{\lambda}=\frac{1}{\lambda} \tag{43}$$

In the CFR model, then, the MTTF and the characteristic life coincide and are equal to $1/\lambda$.

Let us consider, for example, a component with constant failure rate equal to $\lambda = 0.0002$ failures per hour. We want to calculate the MTTF of the component and its reliability after 10000 hours of operation. We'll calculate, then, what is the probability that the component survives other 10000 hours. Assuming, finally, that it has worked without failure for the first 6000 hours, we'll calculate the expected value of the remaining life of the component.

From equation 43 we have:

$$MTTF = \frac{1}{\lambda} = \frac{1}{0.0002\left[\frac{failures}{h}\right]} = 5000\,[h] \tag{44}$$

For the law of the reliability $R(t) = e^{-\lambda \cdot t}$, you get the reliability at 10000 hours:

$$R(10000) = e^{-0.0002 \cdot 10000} = 0.135 \tag{45}$$

The probability that the component survives other 10000 hours, is calculated with the residual reliability. Knowing that this, in the model CFR, is independent from time, we have:

$$R(t + t_0 \mid t_0) = R(t) \rightarrow R(20000 \mid 10000) = R(10000) = 0.135 \tag{46}$$

Suppose now that it has worked without failure for 6000 hours. The expected value of the residual life of the component is calculated using the residual MTTF, that is invariant. In fact:

$$MTTF(t_0) = \int_0^\infty R(t + t_0 \mid t_0) \cdot dt \rightarrow MTTF(6000) = \int_0^\infty R(t + 6000 \mid 6000) \cdot dt = \int_0^\infty R(t) \cdot dt = MTTF = 5000[h] \tag{47}$$

6. CFR in series

Let us consider n different elements, each with its own constant failure rate λ_i and reliability $R_i = e^{-\lambda_i \cdot t}$, arranged in series and let us evaluate the overall reliability R_S. From equation 9 we have:

$$R_S = \prod_{i=1}^n R_i = \prod_{i=1}^n e^{-\lambda_i \cdot t} = e^{-\sum_{i=1}^n \lambda_i \cdot t} \tag{48}$$

Since the reliability of the overall system will take the form of the type $R_S = e^{-\lambda_s \cdot t}$, we can conclude that:

$$R_S = e^{-\sum_{i=1}^n \lambda_i \cdot t} = e^{-\lambda_s \cdot t} \rightarrow \lambda_s = \sum_{i=1}^n \lambda_i \tag{49}$$

In a system of CFR elements arranged in series, then, the failure rate of the system is equal to the sum of failure rates of the components. The MTTF can thus be calculated using the simple relation:

$$MTTF = \frac{1}{\lambda_s} = \frac{1}{\sum\limits_{i=1}^{n} \lambda_i} \tag{50}$$

For example, let me show the following example. A system consists of a pump and a filter, used to separate two parts of a mixture: the concentrate and the squeezing. Knowing that the failure rate of the pump is constant and is $\lambda_P = 1,5 \bullet 10^{-4}$ failures per hour and that the failure rate of the filter is also CFR and is $\lambda_F = 3 \bullet 10^{-5}$, let's try to assess the failure rate of the system, the MTTF and the reliability after one year of continuous operation.

To begin, we compare the physical arrangement with the reliability one, as represented in the following figure:

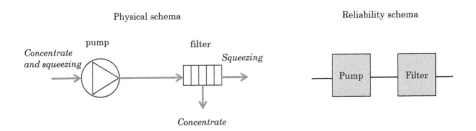

Figure 14. physical and reliability modeling of a pump and a filter producing orange juice.

As can be seen, it is a simple series, for which we can write:

$$\lambda_s = \sum_{i=1}^{n} \lambda_i = \lambda_P + \lambda_F = 1.8 \bullet 10^{-4} \left[\frac{\text{failures}}{\text{h}} \right] \tag{51}$$

$MTTF$ is the reciprocal of the failure rate and can be written:

$$MTTF = \frac{1}{\lambda_s} = \frac{1}{1.8 \bullet 10^{-4}} = 5,555[\text{h}] \tag{52}$$

As a year of continuous operation is $24 \cdot 365 = 8,760$ hours, the reliability after one year is:

$$R_S = e^{-\lambda_s \bullet t} = e^{-1.8 \bullet 10^{-4} \bullet 8760} = 0.2066$$

7. CFR in parallel

If two components arranged in parallel are similar and have constant failure rate λ, the reliability of the system R_P can be calculated with equation 10, wherein R_C is the reliability of the component $R_C = e^{-\lambda t}$:

$$R_P = 1 - \prod_{i=1}^{2}(1 - R_i) = 1 - (1 - R_1)^2 = 2R_C - R_C^2 = 2e^{-\lambda \cdot t} - e^{-2\lambda \cdot t} \tag{53}$$

The calculation of the MTTF leads to $MTTF = \frac{3}{2\lambda}$. In fact we have:

$$MTTF = \int_0^\infty R(t) \cdot dt = \int_0^\infty 2e^{-\lambda \cdot t} - e^{-2\lambda \cdot t} \cdot dt = -\frac{2}{\lambda}e^{-\lambda t} + \frac{1}{2\lambda}e^{-2\lambda t} \Big|_0^\infty = \frac{2}{\lambda}(0 - 1) + \frac{1}{2\lambda}(0 - 1) = \frac{3}{2\lambda} \tag{54}$$

Therefore, the MTTF increases compared to the single component CFR. The failure rate of the parallel system λ_P, reciprocal of the MTTF, is:

$$\lambda_P = \frac{1}{MTTF} = \frac{2}{3}\lambda \tag{55}$$

As you can see, the failure rate is not halved, but was reduced by one third.

For example, let us consider a safety system which consists of two batteries and each one is able to compensate for the lack of electric power of the grid. The two generators are equal and have a constant failure rate $\lambda_B = 9 \cdot 10^{-6}$ failures per hour. We'd like to calculate the failure rate of the system, the MTTF and reliability after one year of continuous operation.

As in the previous case, we start with a reliability block diagram of the problem, as visible in Figure 15.

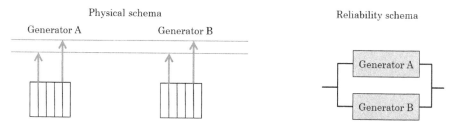

Physical schema   Reliability schema

Generator A   Generator B

Generator A

Generator B

Figure 15. Physical and reliability modeling of an energy supply system.

It is a parallel arrangement, for which the following equation is applicable:

$$\lambda_p = \frac{2}{3}\lambda = \frac{2}{3}9 \cdot 10^{-6} = 6 \cdot 10^{-6}\left[\frac{guasti}{h}\right] \tag{56}$$

The MTTF is the reciprocal of the failure rate and is:

$$MTTF = \frac{1}{\lambda_p} = \frac{1}{6 \cdot 10^{-6}} = 166,666[h] \tag{57}$$

As a year of continuous operation is $24 \cdot 365 = 8,760$ hours, the reliability after one year is:

$$R_p = e^{-\lambda_p \cdot t} = e^{-6 \cdot 10^{-6} \cdot 8,760} = 0.9488 \tag{58}$$

It is interesting to calculate the reliability of a system of identical elements arranged in a parallel configuration k out of n. The system is partially redundant since a group of k elements is able to withstand the load of the system. The reliability is:

$$R_{k \text{ out of } n} = P(k \le j \le n) = \sum_{j=k}^{n} \binom{n}{j} R^j \cdot (1 - R)^{n-j} \tag{59}$$

Let us consider, for example, three electric generators, arranged in parallel and with failure rate $\lambda = 9 \cdot 10^{-6}$. In order for the system to be active, it is sufficient that only two items are in operation. Let's get the reliability after one year of operation.

We'll have: $n = 3$, $k = 2$. So, after a year of operation ($t = 8760\,h$), reliability can be calculated as follows:

$$R_{2 \text{ out of } 3} = \sum_{j=2}^{3} \binom{3}{j} R^j \cdot (1-R)^{3-j} = \binom{3}{2}e^{-2\lambda t}(1 - e^{-\lambda t})^{3-2} + \binom{3}{3}e^{-\lambda t}(1 - e^{-\lambda t})^{3-3} =$$

$$= \frac{3!}{2!(3-2)!}e^{-2\lambda t} \cdot (1 - e^{-\lambda t})^{3-2} + \frac{3!}{3!(3-3)!}e^{-\lambda t} \cdot (1 - e^{-\lambda t})^{3-3} =$$

$$= 3 \cdot e^{-2\lambda t} \cdot (1 - e^{-\lambda t})^{3-2} + 1 \cdot e^{-\lambda t} \cdot (1 - e^{-\lambda t})^{3-3} = 0.963$$

A particular arrangement of components is that of the so-called parallel with stand-by: the second component comes into operation only when the first fails. Otherwise, it is idle.

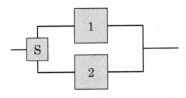

Figure 16. RBD diagram of a parallel system with stand-by. When component 1 fails, the switch S activates component 2. For simplicity, it is assumed that S is not affected by faults.

If the components are similar, then $\lambda_1 = \lambda_2$. It's possible to demonstrate that for the stand-by parallel system we have:

$$MTTF = \frac{2}{\lambda} \tag{60}$$

Thus, in parallel with stand-by, the MTTF is doubled.=

8. Repairable systems

The devices for which it is possible to perform some operations that allow to reactivate the functionality, deserve special attention. A repairable system [6] is a system that, after the failure, can be restored to a functional condition from any action of maintenance, including replacement of the entire system. Maintenance actions performed on a repairable system can be classified into two groups: **Corrective Maintenance - CM** and **Preventive Maintenance - PM**. Corrective maintenance is performed in response to system errors and might correspond to a specific activity of both repair of replacement. Preventive maintenance actions, however, are not performed in response to the failure of the system to repair, but are intended to delay or prevent system failures. Note that the preventive activities are not necessarily cheaper or faster than the corrective actions.

As corrective actions, preventive activities may correspond to both repair and replacement activities. Finally, note that the actions of operational maintenance (servicing) such as, for example, put gas in a vehicle, are not considered PM [7].

Preventative maintenance can be divided into two subcategories: **scheduled** and **on-condition**. Scheduled maintenance (hard-time maintenance) consists of routine maintenance operations, scheduled on the basis of precise measures of elapsed operating time.
Condition-Based Maintenance - CBM [8] (also known as predictive maintenance) is one of the most widely used tools for monitoring of industrial plants and for the management of maintenance policies. The main aim of this approach is to optimize maintenance by reducing costs and increasing availability. In CBM it is necessary to identify, if it exists, a measurable parameter, which expresses, with accuracy, the conditions of degradation of the system. What is needed, therefore, is a physical system of sensors and transducers capable of monitoring the parameter and, thereby, the reliability performance of the plant. The choice of the monitored parameter is crucial, as is its time evolution that lets you know when maintenance action must be undertaken, whether corrective or preventive.

To adopt a CBM policy requires investment in instrumentation and prediction and control systems: you must run a thorough feasibility study to see if the cost of implementing the apparatus are truly sustainable in the system by reducing maintenance costs.

The CBM approach consists of the following steps:

- group the data from the sensors;

- diagnose the condition;
- estimate the Remaining Useful Life – RUL;
- decide whether to maintain or to continue to operate normally.

CBM schedule is modeled with algorithms aiming at high effectiveness, in terms of cost minimization, being subject to constraints such as, for example, the maximum time for the maintenance action, the periods of high production rate, the timing of supply of the pieces parts, the maximization of the availability and so on.

In support of the prognosis, it is now widespread the use of diagrams that do understand, even graphically, when the sensor outputs reach alarm levels. They also set out the alert thresholds that identify ranges of values for which maintenance action must arise [9].

Starting from a state of degradation, detected by a measurement at the time t_k, we calculate the likelihood that the system will still be functioning within the next instant of inspection t_{k+1}. The choice to act with a preventive maintenance is based on the comparison of the expected value of the cost of unavailability, with the costs associated with the repair. Therefore, you create two scenarios:

- continue to operate: if we are in the area of not alarming values. It is also possible that being in the area of preventive maintenance, we opt for a postponement of maintenance because it has already been established replacement intervention within a short interval of time
- stop the task: if we are in the area of values above the threshold established for preventive maintenance of condition.

The modeling of repairable systems is commonly used to evaluate the performance of one or more repairable systems and of the related maintenance policies. The information can also be used in the initial phase of design of the systems themselves.

In the traditional paradigm of modeling, a repairable system can only be in one of two states: working (up) or inoperative (down). Note that a system may not be functioning not only for a fault, but also for preventive or corrective maintenance.

9. Availability

Availability may be generically be defined as the percentage of time that a repairable system is in an operating condition. However, in the literature, there are four specific measures of repairable system availability. We consider only the **limit availability**, defined with the limit of the probability $A(t)$ that the system is working at time t, when t tends to infinity.

$$A = \lim_{t \to \infty} A(t) \tag{61}$$

The limit availability just seen is also called **intrinsic availability**, to distinguish it from the **technical availability**, which also includes the logistics cycle times incidental to maintenance actions (such as waiting for the maintenance, waiting for spare parts, testing...), and from the **operational availability** that encompasses all other factors that contribute to the unavailability of the system such as time of organization and preparation for action in complex and specific business context [10].

The models of the impact of preventive and corrective maintenance on the age of the component, distinguish in perfect, minimal and imperfect maintenance. **Perfect maintenance** (perfect repair) returns the system **as good as new** after maintenance. The **minimal repair**, restores the system to a working condition, but does not reduce the actual age of the system, leaving it **as bad as old**. The imperfect maintenance refers to maintenance actions that have an intermediate impact between the perfect maintenance and minimal repair.

The average duration of maintenance activity is the expected value of the probability distribution of repair time and is called **Mean Time To Repair - MTTR** and is closely connected with the concept of **maintainability**. This consists in the probability of a system, in assigned operating conditions, to be reported in a state in which it can perform the required function.

Figure 17 shows the state functions of two repairable systems with increasing failure rate, maintained with perfect and minimal repair.

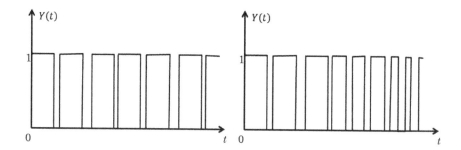

Figure 17. perfect maintenance vs minimal repair. In figure are represented the state functions of two systems both with IFR. $Y(t)$ is equal to 1 when the system wotks, otherwise it's 0. The left system is subject to a policy of perfect repair and shows homogeneous durations of the periods of operation. The right system adopts the minimal repair for which the durations of the periods of operation are reducing as time goes by.

10. The general substitution model

The general substitution model, states that the failure time of a repairable system is an unspecified random variable. The duration of corrective maintenance (perfect) is also a random variable. In this model it is assumed that preventive maintenance is not performed.

Let's denote by T_i the duration of the i - th interval of operation of the repairable system. For the assumption of perfect maintenance (as good as new), $\{T_1, T_2, ..., T_i, ..., T_n\}$ is a sequence of independent and identically distributed random variables.

Let us now designate with D_i the duration of the i - th corrective maintenance action and assume that these random variables are independent and identically distributed. Therefore, each cycle (whether it is an operating cycle or a corrective maintenance action) has an identical probabilistic behavior, and the completion of a maintenance action coincides with time when system state returns operating

Regardless of the probability distributions governing T_i and D_i, the fundamental result of the general pattern of substitution is as follows:

$$A = \frac{E(T_i)}{E(T_i) + E(D_i)} = \frac{MTTF}{MTTF + MTTR} = \frac{MTTF}{MTBF} \tag{62}$$

11. The substitution model for CFR

Let us consider the special case of the general substitution model where T_i is an exponential random variable with constant failure rate λ. Let also D_i be an exponential random variable with constant repair rate μ. Since the reparable system has a constant failure rate (CFR), we know that aging and the impact of corrective maintenance are irrelevant on reliability performance. For this system it can be shown that the limit availability is:

$$A = \frac{\mu}{\lambda + \mu} \tag{63}$$

Let us analyze, for example, a repairable system, subject to a replacement policy, with failure and repair times distributed according to negative exponential distribution. MTTF=1000 hours and MTTR=10 hours.

Let's calculate the limit availability of the system. The formulation of the limit availability in this system is given by eq. 63, so we have:

$$A = \frac{\mu}{\lambda + \mu} = \frac{\frac{1}{10}}{\frac{1}{1000} + \frac{1}{10}} = \frac{0.1}{0.101} = 0.990 \tag{64}$$

This means that the system is available for 99% of the time.

12. General model of minimal repair

After examining the substitution model, we now want to consider a second model for repairable system: the general model of minimal repair. According to this model, the time of system failure is a random variable. Corrective maintenance is instantaneous, the repair is minimal, and not any preventive activity is performed.

The times of arrival of faults, in a repairable system corresponding to the general model of minimal repair, correspond to a process of random experiments, each of which is regulated by the same negative exponential distribution. As known, having neglected the repair time, the number of faults detected by time t, $\{N(t), \ t \geq 0\}$, is a non-homogeneous Poisson process, described by the Poisson distribution.

13. Minimal repair with CFR

A well-known special case of the general model of minimal repair, is obtained if the failure time T is a random variable with exponential distribution, with failure rate λ.

In this case, the general model of minimal repair is simplified because the number $E[N(t)]$ of faults that occur within the time t: $\{N(t), \ t \geq 0\}$ is described by a homogeneous Poisson process with intensity $z(t) = \lambda$, and is:

$$E[N(t)] = \mu_{N(t)} = Z(t) = \int_0^t z(u) \bullet du = \int_0^t \lambda \bullet du = \lambda t \tag{65}$$

If, for example, we consider $\lambda = 0.1$ faults/hour, we obtain the following values at time 100, 1000 and
10000:
$E[N(100)] = 0,1 \bullet 100 = 10$; $E[N(1000)] = 0,1 \bullet 1000 = 100$; $E[N(10000)] = 0,1 \bullet 10000 = 1000$.   It should be noted, as well, a linear trend of the expected number of failures given the width of the interval taken.

Finally, we can obtain the probability mass function of $N(t)$, being a Poisson distribution:

$$P[N(t) = n] = \frac{Z(t)^n}{n!} e^{-Z(t)} = \frac{(\lambda t)^n}{n!} e^{-\lambda t} \tag{66}$$

Also, the probability mass function of $N(t+s) - N(s)$, that is the number of faults in a range of amplitude t shifted forward of s, is identical:

$$P[N(t+s) - N(s) = n] = \frac{(\lambda t)^n}{n!} e^{-\lambda t} \tag{67}$$

Since the two values are equal, the conclusion is that in the homogeneous Poisson process (CFR), the number of faults in a given interval depends only on the range amplitude.

The behavior of a Poisson mass probability distribution, with rate equal to 5 faults each year, representing the probability of having $n \in \mathbf{N}$ faults within a year, is shown in Figure 18.

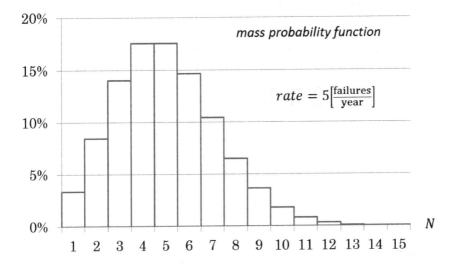

Figure 18. Poisson distribution. In the diagram you can see the probability of having N faults within a year, having a homogeneous Poisson process with a rate of 5 faults each year.

Since in the model of minimal repair with CFR, repair time is supposed to be zero (MTTR = 0), the following relation applies:

$$MTBF = MTTF + MTTR = MTTF = \frac{1}{\lambda} \qquad (68)$$

Suppose that a system, subjected to a repair model of minimal repair, shows failures according to a homogeneous Poisson process with failure rate $\lambda = 0.0025$ failures per hour. We'd like to estimate the average number of failures that the system will have during 5000 hours. Then, determine the probability of having not more than 15 faults in a operation period of 5000 hours.

The estimate of the average number of failures in 5000 hours, can be carried out with the expected value function:

$$E[N(t)] = \lambda \bullet t \rightarrow E[N(5000)] = 0.0025\left[\tfrac{\text{failures}}{\text{h}}\right] \bullet 5000[\text{h}] = 12.5[\text{failures}] \qquad (69)$$

The probability of having not more than 15 faults in a period of 5000 hours of operation, is calculated with the sum of the probability mass function evaluated between 0 and 15:

$$P[N(5000) \leq 15] = \sum_{n=0}^{15} \frac{(\lambda t)^n}{n!} e^{-\lambda \bullet t} = \sum_{n=0}^{15} \frac{12.5^n}{n!} e^{-12.5} = 0.806 \qquad (70)$$

14. Minimal repair: Power law

A second special case of the general model of minimal repair, is obtained if the failure time T is a random variable with a Weibull distribution, with shape parameter β and scale parameter α.

In this case the sequence of failure times is described by a **Non-Homogeneous Poisson Process - NHPP** with intensity $z(t)$ equal to the probability density function of the Weibull distribution:

$$z(t) = \frac{\beta}{\alpha^\beta} t^{\beta-1} \tag{71}$$

Since the cumulative intensity of the process is defined by:

$$Z(t) = \int_0^t z(u) \bullet du \tag{72}$$

the cumulative function is:

$$Z(t) = \int_0^t \frac{\beta}{\alpha^\beta} u^{\beta-1} \bullet du = \frac{\beta}{\alpha^\beta} \bullet \frac{u^\beta}{\beta}\Big|_0^t = \frac{t^\beta}{\alpha^\beta} = \left(\frac{t}{\alpha}\right)^\beta \tag{73}$$

As it can be seen, the average number of faults occurring within the time $t \geq 0$ of this not homogeneous poissonian process $E[N(t)] = Z(t)$, follows the so-called **power law**.
If $\beta > 1$, it means that the intensity function $z(t)$ increases and, being this latter the expression of the average number of failures, it means that faults tend to occur more frequently over time. Conversely, if $\beta < 1$, faults decrease over time.

In fact, if we take $\alpha = 10$ hours ($\lambda = 0.1$ failures/h) and $\beta = 2$, we have: $E[N(100)] = (0.1 \bullet 100)^2 = 100 = 10^2$;　　　　　　　$E[N(1000)] = (0.1 \bullet 1000)^2 = 10000 = 100^2$; $E[N(10000)] = (0.1 \bullet 10000)^2 = 1000000 = 1000^2$. We can observe a trend no longer linear but increasing according to a power law of a multiple of the time width considered.

The probability mass function of $N(t)$ thus becomes:

$$P[N(t) = n] = \frac{Z(t)^n}{n!} e^{-Z(t)} = \frac{\left(\frac{t}{\alpha}\right)^{\beta \bullet n}}{n!} e^{-\left(\frac{t}{\alpha}\right)^\beta} \tag{74}$$

For example, let us consider a system that fails, according to a power law, having $\beta = 2.2$ and $\alpha = 1500$ hours. What is the average number of faults occurring during the first 1000 hours of operation? What is the probability of having two or more failures during the first 1000 hours of operation? Which is the average number of faults in the second 1000 hours of operation?

The average number of failures that occur during the first 1000 hours of operation, is calculated with the expected value of the distribution:

$$E[N(t)] = \mu_{N(t)} = Z(t) = \left(\frac{t}{\alpha}\right)^{\beta} \rightarrow E[N(1000)] = \left(\frac{1000}{1500}\right)^{2.2} = 0.41 \tag{75}$$

The probability of two or more failures during the first 1000 hours of operation can be calculated as complementary to the probability of having zero or one failure:

$$P[N(1000) \geq 2] = 1 - P[N(1000) < 2] = 1 - \sum_{n=0}^{1} \frac{\left(\frac{t}{\alpha}\right)^{\beta \cdot n}}{n!} e^{-\left(\frac{t}{\alpha}\right)^{\beta}} = 1 - \frac{0.41^0}{0!} e^{-0.41} - \frac{0.41^1}{1!} e^{-0.41} = 1 - 0.663 - 0.272 = 0.064 \tag{76}$$

The average number of faults in the succeeding 1000 hours of operation is calculated using the equation:

$$E[N(t+s) - N(s)] = Z(t+s) - Z(s) \tag{77}$$

that, in this case, is:

$$E[N(2000) - N(1000)] = Z(2000) - Z(1000) = 1.47 \tag{78}$$

15. Conclusion

After seeing the main definitions of reliability and maintenance, let's finally see how we can use reliability knowledge also to carry out an economic optimization of replacement activities.

Consider a process that follows the power law with $\beta > 1$. As time goes by, faults begin to take place more frequently and, at some point, it will be convenient to replace the system.

Let us define with τ the time when the replacement (here assumed instantaneous) takes place. We can build a cost model to determine the optimal preventive maintenance time τ^* which optimizes reliability costs.

Let's denote by C_f the cost of a failure and with C_r the cost of replacing the repairable system.

If the repairable system is replaced every τ time units, in that time we will have the replacement costs C_r and so many costs of failure C_f as how many are the expected number of faults in the time range $(0; \tau]$. The latter quantity coincides with the expected value of the number of faults $E[N(\tau)]$.

The average cost per unit of time $c(\tau)$, in the long term, can then be calculated using the following relationship:

$$c(\tau) = \frac{C_f \cdot E[N(\tau)] + C_r}{\tau} \tag{79}$$

Then follows:

$$c(\tau) = \frac{C_f \bullet Z(\tau) + C_r}{\tau} \tag{80}$$

Differentiating $c(\tau)$ with respect to τ and placing the differential equal to zero, we can find the relative minimum of costs, that is, the optimal time τ^* of preventive maintenance. Manipulating algebraically we obtain the following final result:

$$\tau^* = \alpha \bullet \left[\frac{C_r}{C_f (\beta - 1)} \right]^{\frac{1}{\beta}} \tag{81}$$

Consider, for example, a system that fails according to a Weibull distribution with $\beta = 2.2$ and $\alpha = 1500$ hours. Knowing that the system is subject to replacement instantaneous and that the cost of a fault $C_f = 2500$ € and the cost of replacing $C_r = 18000$ €, we want to evaluate the optimal interval of replacement.

The application of eq. 81 provides the answer to the question:

$$\tau^* = \alpha \bullet \left[\frac{C_r}{C_f (\beta - 1)} \right]^{\frac{1}{\beta}} = 1500 \bullet \left[\frac{18000}{2500(2.2 - 1)} \right]^{\frac{1}{2.2}} = 1500 \cdot 2.257 = 3387 \, [h] \tag{82}$$

Nomenclature

RBD: Reliability Block Diagram

CBM: Condition-Based Maintenance

CFR: Constant Failure Rate

CM: Corrective Maintenance

DFR: Decreasing Failure Rate

IFR: Increasing Failure Rate

MCS: Minimal Cut Set

MPS: Minimal Path Set

MTTF: Mean Time To Failure

MTTR: Mean Time To Repair

NHPP: Non-Homogeneous Poisson Process

PM: Preventive Maintenance

Author details

Filippo De Carlo

Address all correspondence to: filippo.decarlo@unifi.it

Industrial Engineering Department, University of Florence, Florence, Italy

References

[1] Nakajima S. Introduction to TPM: Total Productive Maintenance. Productivity Press, Inc., 1988, 1988:129.

[2] Barlow RE. Engineering Reliability. SIAM; 2003.

[3] De Carlo F. Impianti industriali: conoscere e progettare i sistemi produttivi. New York: Mario Tucci; 2012.

[4] O'Connor P, Kleyner A. Practical Reliability Engineering. John Wiley & Sons; 2011.

[5] Meyer P. Understanding Measurement: Reliability. Oxford University Press; 2010.

[6] Ascher H, Feingold H. Repairable systems reliability: modeling, inference, misconceptions and their causes. M. Dekker; 1984.

[7] De Carlo F, Borgia O, Adriani PG, Paoli M. New maintenance opportunities in legacy plants. 34th ESReDA Seminar, San Sebastian, Spain: 2008.

[8] Gertler J. Fault detection and diagnosis in engineering systems. Marcel Dekker; 1998.

[9] Borgia O, De Carlo F, Tucci M. From diagnosis to prognosis: A maintenance experience for an electric locomotive. Safety, Reliability and Risk Analysis: Theory, Methods and Applications - Proceedings of the Joint ESREL and SRA-Europe Conference, vol. 1, 2009, pp. 211–8.

[10] Racioppi G, Monaci G, Michelassi C, Saccardi D, Borgia O, De Carlo F. Availability assessment for a gas plant. Petroleum Technology Quarterly 2008;13:33–7.

Chapter 5

Production Scheduling Approaches for Operations Management

Marcello Fera, Fabio Fruggiero, Alfredo Lambiase,
Giada Martino and Maria Elena Nenni

Additional information is available at the end of the chapter

1. Introduction

Scheduling is essentially the short-term execution plan of a production planning model. Production scheduling consists of the activities performed in a manufacturing company in order to manage and control the execution of a production process. A schedule is an assignment problem that describes into details (in terms of minutes or seconds) which activities must be performed and how the factory's resources should be utilized to satisfy the plan. Detailed scheduling is essentially the problem of allocating machines to competing jobs over time, subject to the constraints. Each work center can process one job at a time and each machine can handle at most one task at a time. A scheduling problem, typically, assumes a fixed number of jobs and each job has its own parameters (i.e., tasks, the necessary sequential constraints, the time estimates for each operation and the required resources, no cancellations). All scheduling approaches require some estimate of how long it takes to perform the work. Scheduling affects, and is affected by, the shop floor organization. All scheduling changes can be projected over time enabling the identification and analysis of starting time, completion times, idle time of resources, lateness, etc.…

A right scheduling plan can drive the forecast to anticipate completion date for each released part and to provide data for deciding what to work on next. Questions about "Can we do it?" and/or "How are we doing?" presume the existence of approaches for optimisation. The aim of a scheduling study is, in general, to perform the tasks in order to comply with priority rules and to respond to strategy. An optimal short-term production planning model aims at gaining time and saving opportunities. It starts from the execution orders and it tries to allocate, in the best possible way, the production of the different items to the facilities. A good schedule starts from planning and springs from respecting resource conflicts, managing the release of jobs to a shop and optimizing completion time of all jobs. It defines the starting time of each task and determines whatever and how delivery promises can be met. The minimization of one or more objectives has to be accomplished (e.g., the number of jobs that are shipped late, the minimization set up costs, the maximum completion time of jobs, maximization of throughput, etc.). Criteria could be ranked from applying simple rules to determine which job has to be processed next at which work-centre (i.e., dispatching) or to the use of advanced optimizing methods that try to maximize the performance of the given environment. Fortunately many of these objectives are mutually supportive (e.g., reducing manufacturing lead time reduces work in process and increases probability to meeting due dates). To identify the exact sequence among a plethora of possible combinations, the final schedule needs to apply rules in order to quantify urgency of each order (e.g., assigned order's due date - defined as global exploited strategy; amount of processing that each order requires - generally the basis of a local visibility strategy). It's up to operations management to optimize the use of limited resources. Rules combined into *heuristic*[1] approaches and, more in general, in upper level multi-objective methodologies (i.e., *meta-heuristics*[2]), become the only methods for scheduling when dimension and/or complexity of the problem is outstanding [1]. In the past few years, metaheuristics have received much attention from the hard optimization community as a powerful tool, since they have been demonstrating very promising results from experimentation and practices in many engineering areas. Therefore, many recent researches on scheduling problems focused on these techniques. Mathematical analyses of metaheuristics have been presented in literature [2, 3].

This research examines the main characteristics of the most promising meta-heuristic approaches for the general process of a Job Shop Scheduling Problems (i.e., JSSP). Being a NP complete and highly constrained problem, the resolution of the JSSP is recognized as a key point for the factory optimization process [4]. The chapter examines the soundness and key contributions of the 7 meta-heuristics (i.e., Genetics Approaches, Ants Colony Optimization, Bees Algorithm, Electromagnetic Like Algorithm, Simulating Annealing, Tabu Search and Neural Networks), those that improved the production scheduling vision. It reviews their accomplishments and it discusses the perspectives of each meta approach. The work represents a practitioner guide to the implementation of these meta-heuristics in scheduling job shop processes. It focuses on the logic, the parameters, representation schemata and operators they need.

2. The job shop scheduling problem

The two key problems in production scheduling are „priorities" and „capacity". Wight (1974) described *scheduling* as „establishing the timing for performing a task" and observes that, in

---

1 The etymology of the word heuristic derives from a Greek word *heurìsco (εὑρίσκω)* - it means „to find"- and is considered the art of discovering new strategy rules to solve problems. Heuristics aims at a solution that is „good enough" in a computing time that is „small enough".

2 The term metaheuristc originates from union of prefix *meta (μετα)* - it means „behind, in the sense upper level methodology" – and word *heuristic* - it means „to find". Metaheuristcs' search methods can be defined as upper level general methodologies guiding strategies in designing heuristics to obtain optimisation in problems.

manufacturing firms, there are multiple types of scheduling, including the detailed scheduling of a shop order that shows when each operation must start and be completed [5]. Baker (1974) defined scheduling as „a plan than usually tells us when things are supposed to happen" [6]. Cox *et al.* (1992) defined *detailed scheduling* as „the actual assignment of starting and/or completion dates to operations or groups of operations to show when these must be done if the manufacturing order is to be completed on time"[7]. Pinedo (1995) listed a number of important surveys on production scheduling [8]. For Hopp and Spearman (1996) „scheduling is the allocation of shared resources over time to competing activities" [9]. Makowitz and Wein (2001) classified production scheduling problems based on attributes: the presence of setups, the presence of due dates, the type of products.

Practical scheduling problems, although more highly constrained, are high difficult to solve due to the number and variety of jobs, tasks and potentially conflicting goals. Recently, a lot of Advanced Production Scheduling tools arose into the market (e.g., Aspen PlantTM Scheduler family, Asprova, R2T – Resourse To Time, DS APS – DemandSolutions APS, DMS – Dynafact Manufacturing System, i68Group, ICRON-APS, JobPack, iFRP, Infor SCM, SchveduelePro, Optiflow-Le, Production One APS, MQM – Machine Queue Management, MOM4, JDA software, Rob-ex, Schedlyzer, OMP Plus, MLS and MLP, Oracle Advanced Scheduling, Ortec Schedule, ORTEMS Productionscheduler, Outperform, AIMMS, Planet Together, Preactor, Quintiq, FactoryTalk Scheduler, SAP APO-PP/DS, and others). Each of these automatically reports graphs. Their goal is to drive the scheduling for assigned manufacturing processes. They implement rules and optimise an isolated sub-problem but none of the them will optimise a multi stage resource assignment and sequencing problem.

In a Job Shop (i.e., JS) problem a classic and most general factory environment, different tasks or operations must be performed to complete a job [10]; moreover, priorities and capacity problems are faced for different jobs, multiple tasks and different routes. In this contest, each job has its own individual flow pattern through assigned machines, each machine can process only one operation at a time and each operation can be processed by only one machine at a time. The purpose of the procedure is to obtain a schedule which aims to complete all jobs and, at the same time, to minimize (or maximize) the objective function. Mathematically, the JS Scheduling Problem (i.e., JSSP) can be characterized as a combinatorial optimization problem. It has been generally shown to be NP-hard[3] belonging to the most intractable problems considered [4, 11, 12]. This means that the computation effort may grow too fast and there are not universal methods making it possible to solve all the cases effectively. Just to understand what the technical term means, consider the single-machine sequencing problem with three jobs. How many ways of sequencing three jobs do exist? Only one of the three jobs could be in the first position, which leaves two candidates for the second position and only one for the last position. Therefore the no. of permutations is 3!. Thus, if we want to optimize, we need to consider six alternatives. This means that as the no. of jobs to be sequenced becomes larger (i.e., $n>80$), the no. of possible sequences become quite ominous and an exponential function dominates the amount of time required to find the optimal solution [13]. Scheduling, however,

---

3 A problem is NP-complete if exists no algorithm that solves the problem in a polynomial time. A problem is NP-hard if it is possible to show that it can solve a NP-complete problem.

performs the definition of the optimal sequence of n jobs in m machines. If a set of n jobs is to be scheduled on m machines, there are $(n!)^m$ possible ways to schedule the job.

It has to undergo a discrete number of operations (i.e., *tasks*) on different resources (i.e., *machines*). Each product has a fixed route defined in the planning phase and following processing requirements (i.e., precedence constraints). Other constraints, e.g. zoning which binds the assignment of task to fixed resource, are also taken into consideration. Each machine can process only one operation at a time with no interruptions (pre-emption). The schedule we must derive aims to complete all jobs with minimization (maximization) of an objective function on the given production plant.

Let:

- $J = \{J_1, J_2, \ldots\ldots, J_n\}$ the set of the job order existing inside the system;

- $M = \{M_1, M_2, \ldots\ldots, M_m\}$ the set of machines that make up the system.

JSSP, marked as Π_j, consists in a finite set J of n jobs $\{J_i\}_{i=1}^n$. Each J_i is characterized by a manufacturing cycle CL_i regarded as a finite set M of m machines $\{M_k\}_{k=1}^m$ with an uninterrupted processing time τ_{ik}. J_i, $\forall\ i = 1, \ldots, n$, is processed on a fixed machine m_i and requires a chain of tasks $O_{i1}, O_{i2}, \ldots\ldots, O_{im_i}$, scheduled under precedence constraints. O_{ik} is the task of job J_i which has to be processed on machine M_k for an uninterrupted processing time period τ_{ik} and no operations may pre-empted.

To accommodate extreme variability in different parts of a job shop, schedulers separate workloads in each work-centres rather than aggregating them [14]. Of more than 100 different rules proposed by researchers and applied by practitioners exist, some have become common in Operations Management systems: First come- First served, Shortest Processing Time, Earliest Due Date, Slack Time Remaining, Slack Time Remaining For each Operation, Critical Ratio, Operation Due Date, etc. [15]. Besides these, Makespan is often the performance feature in the study of resource allocation [16]. Makespan represents the time elapsed from the start of the first task to the end of the last task in schedule. The minimisation of makespan arranges tasks in order to level the differences between the completion time of each work phase. It tries to smooth picks in work-centre occupancy to obtain batching in load assignment per time. Although direct time constraints, such as minimization of processing time or earliest due date, are sufficient to optimize industrial scheduling problems, for the reasons as above the minimization of the makespan is preferable for general/global optimization performances because it enhances the overall efficiency in shop floor and reduces manufacturing lead time variability [17].

Thus, in JSSP optimization variant of Π_j, the objective of a scheduling problem is typically to assign the tasks to time intervals in order to minimise the makespan and referred to as:

$$c'_{max}(t) = f(CL_i, \tau_{ik}, s_{ik}), \forall\ i = 1\ldots n; \forall\ k = 1\ldots m \tag{1}$$

where t represent time (i.e. iteration steps)

$$C'_{max}(t) = \min(C_{max}(t)) = \min\{\max_i[\tau_{i k} + s_{i k}] : \forall J_i \in J, \forall M_k \in M\} \quad (2)$$

and $s_{i k} \geq 0$ represents the starting time of k-th operation of i-th job. $s_{i k}$ is the time value that we would like to determinate in order to establish the suited schedule activities order.

3. Representation of scheduling instances

The possible representation of a JS problem could be done through a Gantt chart or through a Network representation.

Gantt (1916) created innovative charts for visualizing planned and actual production [18]. According to Cox et al. (1992), a *Gantt chart* is „the earliest and best known type of control chart especially designed to show graphically the relationship between planned performance and actual performance" [19]. Gantt designed his charts so that foremen or other supervisors could quickly know whether production was on schedule, ahead of schedule or behind schedule. A Gantt chart, or bar chart as it is usually named, measures activities by the amount of time needed to complete them and use the space on the chart to represent the amount of the activity that should have been done in that time [7].

A Network representation was first introduced by Roy and Sussman [20]. The representation is based on *"disjunctive graph model"* [21]. This representation starts from the concept that a feasible and optimal solution of JSP can originate from a permutation of task's order. Tasks are defined in a network representation through a probabilistic model, observing the precedence constraints, characterized in a machine occupation matrix M and considering the processing time of each tasks, defined in a time occupation matrix T.

$$M = \begin{pmatrix} M_{11} & \cdots & M_{1n} \\ \vdots & \ddots & \vdots \\ M_{n1} & \cdots & M_{nn} \end{pmatrix} ; \quad T = \begin{pmatrix} \tau(M_{11}) & \cdots & \tau(M_{1n}) \\ \vdots & \ddots & \vdots \\ \tau(M_{n1}) & \cdots & \tau(M_{nn}) \end{pmatrix}$$

JS processes are mathematically described as disjunctive graph $G = (V, C, E)$. The descriptions and notations as follow are due to Adams et. al. [22], where:

- V is a set of nodes representing tasks of jobs. Two additional dummy tasks are to be considered: a *source(0)* node and a *sink(*)* node which stand respectively for the Source (S) task $\tau_0 = 0$, necessary to specify which job will be scheduled first, and an end fixed sink where schedule ends (T) $\tau_* = 0$;

- C is the set of conjunctive arcs or direct arcs that connect two consecutive tasks belonging to the same job chain. These represent technological sequences of machines for each job;

- $E = \bigcup_{r=1}^{m} D_r$, where D_r is a set of disjunctive arcs or not-direct arcs representing pair of operations that must be performed on the same machine M_r.

Each job-tasks pair (i,j) is to be processed on a specified machine $M(i,j)$ for $T(i,j)$ time units, so each node of graph is weighted with j operation's processing time. In this representation all nodes are weighted with exception of source and sink node. This procedure makes always available feasible schedules which don't violate hard constraints[4]. A graph representation of a simple instance of JSP, consisting of 9 operations partitioned into 3 jobs and 3 machines, is presented in fig. 1. Here the nodes correspond to operations numbered with consecutive ordinal values adding two fictitious additional ones:S = "source node" and T = "sink node". The processing time for each operation is the weighted value τ_{ij} attached to the corresponding node,$v \in V$, and for the special nodes, $\tau_0 = \tau_* = 0$.

Let s_v be the starting time of an operation to a node v. By using the disjunctive graph notation, the JSPP can be formulated as a mathematical programming model as follows:

Minimize s_* subject to:

$$s_w - s_v \geq \tau_v \; (v, \, w) \in C \tag{3}$$

$$s_v \geq 0 v \in V \tag{4}$$

$$s_w - s_v \geq \tau_v \vee s_v - s_w \geq \tau_w (v, \; w) \in D_r, 1 \leq r \leq m, \tag{5}$$

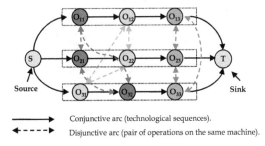

────────▶   Conjunctive arc (technological sequences).

◀─ ─ ─ ─▶   Disjunctive arc (pair of operations on the same machine).

Figure 1. Disjunctive graph representation. There are disjunctive arcs between every pair of tasks that has to be processed on the same machine (dashed lines) and conjunctive arcs between every pair of tasks that are in the same job (dotted lines). Omitting processing time, the problem specification is $O = \{o_{ij}, (i, \; j) \in \{1, 2, 3\}^2\}$, $J = \{J_i = \{o_{ij}\}, (i, \; j) = 1, 2, 3\}$, $M = \{M_j = \{o_{ij}\}, (i, \; j) = 1, 2, 3\}$. Job notation is used.

---

4 Hard constraints are physical ones, while soft constraints are generally those related to human factor e.g., relaxation, fatigue etc…

s_s is equal to the completion time of the last operation of the schedule, which is therefore equal to C_{max}. The first inequality ensures that when there is a conjunctive arc from a node v to a node w, w must wait of least τ_v time after v is started, so that the predefined technological constraints about sequence of machines for each job is not violated. The second condition ensures time to start continuities. The third condition affirms that, when there is a disjunctive arc between a node v and a node w, one has to select either v to be processed prior to w (and w waits for at least τ_v time period) or the other way around, this avoids overlap in time due to contemporaneous operations on the same machine.

In order to obtain a scheduling solution and to evaluate makespan, we have to collect all feasible permutations of tasks to transform the undirected arcs in directed ones in such a way that there are no cycles.

The total number of nodes, $n = (|O| + 2)$ - fixed by taking into account the total number of tasks $|O|$, is properly the total number of operations with more two fictitious ones. While the total number of arcs, in job notation, is fixed considering the number of tasks and jobs of instance:

$$\text{arcs} = \binom{n}{2} + 2 \times |J| = \frac{(|O|) \times \{(|O|) - 1\}}{2} + 2 \times |J| \qquad (6)$$

The number of arcs defines the possible combination paths. Each path from source to sink is a candidate solution for JSSP. The routing graph is reported in figure 2:

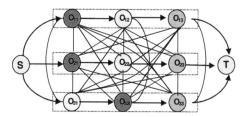

Figure 2. Problem routing representation.

4. Meta-heuristics for solving the JSSP

A logic has to be implemented in order to translate the scheduling problem into an algorithm structure. Academic researches on scheduling problems have produced countless papers [23]. Scheduling has been faced from many perspectives, using formulations and tools of various disciplines such as control theory, physical science and artificial intelligence systems [24]. Criteria for optimization could be ranked from applying simple priority rules to determine which job has to be processed next at the work-centres (i.e., dispatching) to the use of advanced optimizing methods that try to maximize the performance of the given environment [25]. Their way to solution is generally approximate – heuristics – but it constitutes promising alternatives to the exact methods and becomes the only one possible when dimension and/or complexity of the problem is outstanding [26].

Guidelines in using heuristics in combinatorial optimization can be found in Hertz (2003) [27]. A classification of heuristic methods was proposed by Zanakis et al. (1989) [28]. Heuristics are generally classified into *constructive heuristics* and *improvement heuristics*. The first ones are focused on producing a solution based on an initial proposal, the goal is to decrease the solution until all the jobs are assigned to a machine, not considering the size of the problem [29]. The second ones are iterative algorithms which explore solutions by moving step by step form one solution to another. The method starts with an arbitrary solution and transits from one solution to another according to a series of basic modifications defined on case by case basis [30].

Relatively simple rules in guiding heuristic, with exploitation and exploration, are capable to produce better quality solutions than other algorithms from the literature for some classes of instances. These variants originate the class of meta-heuristic approaches [31]. The meta-heuristics[5], and in general the heuristics, do not ensure optimal results but they usually tend to work well [32]. The purpose of the paper is to illustrate the most promising optimization methods for the JSSP.

As optimization techniques, metaheuristics are stochastic algorithms aiming to solve a broad range of hard optimization problems, for which one does not know more effective traditional methods. Often inspired by analogies with reality, such as physics science, Simulated Annealing [33] and Electromagnetic like Methods [34], biology (Genetic Algorithms [35], Tabu Search [36]) and ethnology (Ant Colony [37,], Bees Algorithm [38]), human science (Neural Networks [39]), they are generally of discrete origin but can be adapted to the other types of problems.

4.1. Genetic Algorithms (GAs)

The methodology of a GAs - based on the evolutionary strategy- trasforms a population (set) of individual objects, each with an associated *fitness* value, into a new *generation* of the population occurring genetic operations such as *crossover* (*sexual recombination*) and *mutation* (fig. 3).

The theory of evolutionary computing was formalized by Holland in 1975 [40]. GAs are stochastic search procedures for combinatorial optimization problems based on Darwinian principle of natural reproduction, survival and environment's adaptability [41]. The theory of evolution is biologically explained, the individuals with a stronger fitness are considered better able to survive.. Cells, with one or more strings of DNA (i.e., a chromosome), make up an individual. The gene (i.e., a bit of chromosome located into its particular locus) is, responsible for encoding traits (i.e., alleles). Physical manifestations are raised into genotype (i.e., disposition of genes). Each genotype has is physical manifestation into phenotype. According to these parameters is possible to define a fitness value. Combining individuals through a

---

5 The term metaheuristics was introduced by F. Glover in the paper about Tabu search.

crossover (i.e., recombination of genetic characteristics of parents) across the sexual reproduction, the chromosomal inheritance process performs to offspring. In each epoch a stochastic mutation procedure occurs. The implemented algorithm is able to simulate the natural process of evolution, coupling solution of scheduling route in order to determinate an optimal tasks assignment. Generally, GA has different basic component: representation, initial population, evaluation function, the reproduction selection scheme, genetic operators (mutation and crossover) and stopping criteria. Central to success of any GA is the suitability of its representation to the problem at hand [42]. This is the encoding from the solution of the problem domain to the genetic representation.

During the last decades, different representation's schemata for JS have been proposed, such as *permutation with repetition*. It uses sequence of repeated jobs identifier (e.g., its corresponding cardinal number) to represent solutions [43]. According to the instance in issue, each of the N jobs identifiers will be repeated M times, once for each task. The first time that job's identifier, reading from left to right, will appear means the first task of that job. In this way, precedence constraints are satisfied. The redundancy is the most common caveat of this representation. A proposal of permutation with repetition applying a Generalized Order crossover (GOX) with band |2 3 1 1| of parent 1 moves from PARENT1 [3 2 3 1 1 1 3 2 2] and PARENT2 [2 3 2 1 3 3 2 1 1] to CHILD1 [2 3 1 1 3 2 3 2 1] and CHILD2 [3 2 1 3 2 1 1 3 2].

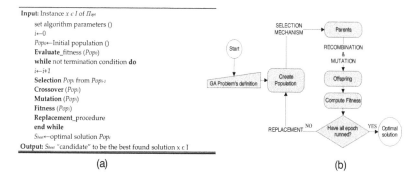

Figure 3. The Genetic Algorithms (GAs) model; 3a. the pseudo-code of a GA; 3b. the flow chart of a general GA.

A mutation operator is applied changing the genes into the same genotype (in order to generate only feasible solutions, i.e., without the rejection procedure). Mutation allows to diversify the search over a broader solution domain and it is needed when there is low level of crossover. Among solutions, the allocation with favourable fitness will have higher probability to be selected through the selection mechanisms.

Another important issue for the GA is the selection mechanism (e.g., Tournament Selection procedure and Roulette Wheel as commonly used [44] - their performances are quite similar attending in the convergence time). The *tournament selection* procedure is based on analogy with competition field, between the genotypes in tournament, the individual which will win (e.g., the one with the best fitness value) is placed in the mating pool. Likewise, in the *roulette wheel selection* mechanism each individual of population has a selection's likelihood proportional to its objective score (in analogy with the real roulette item) and with a probability equal to one of a ball in a roulette, one of the solutions is chosen.

It is very important, for the GAs success, to select the correct ratio between crossover and mutation, because the first one allows to allows to diversify a search field, while a mutation to modify a solution.

4.2. Ant Colony Optimization (ACO) algorithms

If we are on a pic-nic and peer into our cake bitten by a colony of ants, moving in a tidy way and caring on a lay-out that is the optimal one in view of stumbling-blocks and length, we discover how remarkable is nature and we find its evolution as the inspiring source for investigations on intelligence operation scheduling techniques [45]. Natural ants are capable to establish the shortest route path from their colony to feeding sources, relying on the phenomena of *swarm intelligence* for survival. They make decisions that seemingly require an high degree of co-operation, smelling and following a chemical substance (i.e. pheromone[6]) laid on the ground and proportional to goodness load that they carry on (i.e. in a scheduling approach, the goodness of the objective function, reported to makespan in this applicative case).

The same behaviour of natural ants can be overcome in an artificial system with an artificial communication strategy regard as a direct metaphoric representation of natural evolution. The essential idea of an ACO model is that „good solutions are not the result of a sporadic good approach to the problem but the incremental output of good partial solutions item. Artificial ants are quite different by their natural progenitors, maintaining a memory of the step before the last one [37]. Computationally, ACO [46] are population based approach built on stochastic solution construction procedures with a retroactive control improvement, that build solution route with a probabilistic approach and through a suitable selection procedure by taking into account: *(a) heuristic information* on the problem instance being solved; *(b)* (mat-made) *pheromone amount*, different from ant to ant, which stores up and evaporates dynamically at run-time to reflect the agents' acquired search training and elapsed time factor.

The initial schedule is constructed by taking into account heuristic information, initial pheromone setting and, if several routes are applicable, a self-created selection procedure chooses the task to process. The same process is followed during the whole run time. The probabilistic approach focused on pheromone. Path's attractive raises with path choice and probability increases with the number of times that the same path was chosen before [47]. At the same time, the employment of heuristic information can guide the ants towards the most promising solutions and additionally, the use of an agent's colony can give the algorithm: *(i)* Robustness on a fixed solution; *(ii)* Flexibility between different paths.

The approach focuses on co-operative ant colony food retrieval applied to scheduling routing problems. Colorni et al, basing on studies of Dorigo *et al.* [48], were the first to apply Ant System

---

6 It is an organic compound highly volatile that shares on central neural system as an actions' releaser.

(AS) to job scheduling problem [49] and dubbed this approach as **A**nt **C**olony **O**ptimization (ACO). They iteratively create route, adding components to partial solution, by taking into account heuristic information on the problem instance being solved (i.e. visibility) and "artificial" pheromone trials (with its storing and evaporation criteria). Across the representation of scheduling problem like acyclic graph, see fig. 2, the ant's rooting from source to food is assimilated to the scheduling sequence. Think at ants as agents, nodes like tasks and arcs as the release of production order. According to constraints, the ants perform a path from the row material warehouse to the final products one.

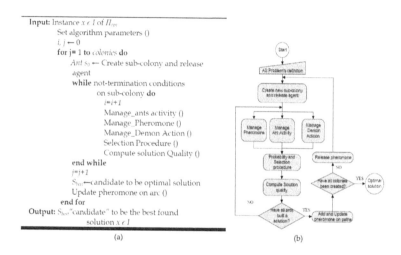

(a)   (b)

Figure 4. The Ant Colony Optimization (ACO) model; 4a. the pseudo-code of an ACO algorithm; 4b. the flow chart of a general ACO procedure.

Constraints are introduced hanging from jobs and resources. Fitness is introduced to translate how good the explored route was. Artificial ants live in a computer realized world. They have an overview of the problem instance they are going to solve across a visibility factor. In the Job Shop side of ACO implementation the visibility has chosen tied with the run time of the task (Eq. 7). The information was about the inquired task's (i.e., j) completion time $Ctime_j$ and idle time $Itime_j$ from the previous position (i.e., i):

$$\eta_{ij}(t) = \frac{1}{(Ctime_j - Itime_j)} = \frac{1}{Rtime_j} \tag{7}$$

The colony is composed of a fixed number of agents $ant=1,\ldots, n$. A probability is associated to each feasible movement ($S_{ant}(t)$) and a selection procedure (generally based on *RWS* or *Tournament* procedure) is applied.

$0 \leq P_{ij}^{ant}(t) \leq 1$ is the probability that at time t the generic agent ant chooses edge $i \rightarrow j$ as next routing path; at time t each ant chooses the next operation where it will be at time $t+1$. This value is valuated through visibility (η) and pheromone (τ) information. The probability value (Eq. 8) is associated to a fitness into selection step.

$$P_{ij}^{ant}(t) = \begin{cases} \dfrac{[\tau_{ij}(t)]^{\alpha} [\eta_{ij}(t)]^{\beta}}{\displaystyle\sum_{h \in S_{ant}(t)} [\tau_{ih}(t)]^{\alpha} [\eta_{ih}(t)]^{\beta}}_{i} & \text{if } j \in S_{ant}(t) \\ 0 & \text{otherwise} \end{cases} \tag{8}$$

Where: $\tau_{ij}(t)$ represents the intensity of trail on connection (i, j) at time t. Set the intensity of pheromone at iteration $t=0$: $\tau_{ij}(o)$ to a general small positive constant in order to ensure the avoiding of local optimal solution; α and β are user's defined values tuning the relative importance of the pheromone vs. the heuristic time-distance coefficient. They have to be chosen $0 < \alpha, \beta \leq 10$ (in order to assure a right selection pressure).

For each cycle the agents of the colony are going out of source in search of food. When all colony agents have constructed a complete path, i.e. the sequence of feasible order of visited nodes, a pheromone update rule is applied (Eq. 9):

$$\tau_{ij}(t + 1) = (1 - \lambda)\tau_{ij}(t + 1) + \Delta\tau_{ij}(t) \tag{9}$$

Besides ants' activity, *pheromone trail evaporation* has been included trough a coefficient representing pheromone vanishing during elapsing time. These parameters imitate the natural world decreasing of pheromone trail intensity over time. It implements a useful form of forgetting. It has been considered a simple decay coefficient (i.e., $0 < \lambda < 1$) that works on total laid pheromone level between time t and $t+1$.

The laid pheromone on the inquired path is evaluated taking into consideration how many agents chose that path and how was the objective value of that path (Eq. 10). The weight of the solution goodness is the makespan (i.e., L_{ant}). A constant of pheromone updating (i.e., Q), equal for all ants and user, defined according to the tuning of the algorithm, is introduced as quantity of pheromone per unit of time (Eq. 11). The algorithm works as follow. It is computed the makespan value for each agent of the colony ($L_{ant}(0)$), following visibility and pheromone defined initially by the user ($\tau_{ij}(0)$) equal for all connections. It is evaluated and laid, according to the disjunctive graph representation of the instance in issue, the amount of pheromone on each arc (evaporation coefficient is applied to design the environment at the next step).

$$\Delta\tau_{ij}(t) = \sum_{ant=1}^{ants} \Delta\tau_{ij}^{ant}(t) \tag{10}$$

$$\Delta\tau_{ij}^{ant}(t) = \begin{cases} \dfrac{Q}{L_{ant}} & \text{if ant-th followed edge (i,j)} \\ 0 & \text{otherwise} \end{cases} \tag{11}$$

Visibility and updated pheromone trail fixes the probability (i.e., the fitness values) of each node (i.e., task) at each iteration; for each cycle, it is evaluated the output of the objective function ($L_{ant}(t)$). An objective function value is optimised accordingly to partial good solution. In this improvement, relative importance is given to the parameters α and β. Good elements for choosing these two parameters are: $\alpha/\beta \cong 0$ (which means low level of α) and little value of α ($0 < \alpha \leq 2$) while ranging β in a larger range ($0 < \beta \leq 6$).

4.3. Bees Algorithm (BA) approach

A colony of bees exploits, in multiple directions simultaneously, food sources in the form of antera with plentiful amounts of nectar or pollen. They are able to cover kilometric distances for good foraging fields [50]. Flower paths are covered based on a stigmergic approach – more nectar places should be visited by more bees [51].

The foraging strategies in colonies of bees starts by scout bees – a percentage of beehive population. They wave randomly from one patch to another. Returning at the hive, those scout bees deposit their nectar or polled and start a recruiting mechanism rated above a certain quality threshold on nectar stored [52]. The recruiting mechanism is properly a launching into a wild dance over the honeycomb. This natural process is known as waggle dance" [53]. Bees, stirring up for discovery, flutter in a number from one to one hundred circuits with a waving and returning phase. The waving phase contains information about direction and distance of flower patches. Waving phases in ascending order on vertical honeycomb suggest flower patches on straightforward line with sunbeams. This information is passed using a kind of dance, that is possible to be developed on right or on left. So through this dance, it is possible to understand the distance from the flower, the presence of nectar and the sunbeam side to choose [54].

The waggle dance is used as a guide or a map to evaluate merits of explored different patches and to exploit better solutions. After waggle dancing on the dance floor, the dancer (i.e. the scout bee) goes back to the flower patch with follower bees that were waiting inside the hive. A squadron moves forward into the patches. More follower bees are sent to more promising patches, while harvest paths are explored but they are not carried out in the long term. A swarm intelligent approach is constituted [55]. This allows the colony to gather food quickly and efficiently with a recursive recruiting mechanism [56].

The Bees Algorithm (i.e., BA) is a population-based search; it is inspired to this natural process [38]. In its basic version, the algorithm performs a kind of neighbourhood search combined with random search. Advanced mechanisms could be guided by genetics [57] or taboo operators [58]. The standard Bees Algorithm first developed in Pham and Karaboga in 2006 [59, 60] requires a set of parameters: no. of scout bees (n), no. of sites selected out of n visited sites (m), no. of best sites out of m selected sites (e), no. of bees recruited for the best e sites (nep), no. of bees recruited for the other *m-e* selected sites (*nsp*), initial size of patches (*ngh*). The standard BA starts with random search.

The honey bees' effective foraging strategy can be applied in operation management problems such as JSSP. For each solution, a complete schedule of operations in JSP is produced. The makespan of the solution is analogous to the profitability of the food source in terms of distance and sweetness of the nectar. Bees, *n* scouts, explore patches, *m* sites - initially a scout bee for each path could be set, over total ways, *ngh*, accordingly to the disjunctive graph of fig. 2, randomly at the first stage, choosing the shorter makespan and the higher profitability of the solution path after the first iteration.

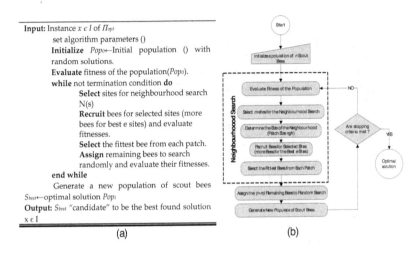

Input: Instance $x \in I$ of Π_{opt}
  set algorithm parameters ()
  Initialize $Pop_0 \leftarrow$ Initial population () with random solutions.
  Evaluate fitness of the population(Pop_0).
  while not termination condition **do**
    Select sites for neighbourhood search N(s)
    Recruit bees for selected sites (more bees for best e sites) and evaluate fitnesses.
    Select the fittest bee from each patch.
    Assign remaining bees to search randomly and evaluate their fitnesses.
  end while
    Generate a new population of scout bees
  $S_{best} \leftarrow$ optimal solution Pop_i
  Output: S_{best} "candidate" to be the best found solution $x \in I$

(a)   (b)

Figure 5. The Bees Algorithm model; 6a. the BA pseudo code; 6b. the flow chart of a general BA procedure.

Together with scouting, this differential recruitment is the key operation of the BA. Once a feasible solution is found, each bee will return to the hive to perform a waggle dance. The output of the waggle dance will be represented by a list of "elite solutions", *e* best selected sites, from which recruited bees, *nep*, are chosen for exploration from the population into the hive. Researches of patches are conducted, other *nsp* bees, in the neighbourhood of the selected sites, *m-e* sites. System maintains, step repetition: *imax*, where each bee of the colony of bees will traverse a potential solution. Flower patches, *e* sites, with better fitness (makespan) will have a higher probability for "elite solutions", promoting the exploitation to an optimal solution.

4.4. Electromagnetism like Method (EM)

The Electromagnetic Like Algorithm is a population based meta-heuristics proposed by Birbil and Fang [61] to tackle with combinatorial optimisation problems. Algorithm is based on the natural law of attraction and repulsion between charges (Coulomb's law) [62]. EM simulates electromagnetic interaction [63]. The algorithm evaluates fitness of solutions considering charge of particles. Each particle represents a solution. Two points into the space had different charges in relation to what electromagnetic field acts on them [64]. An electrostatic force, in repulsion or attraction, manifests between two points charges. The electrostatic force is directly proportional to the magnitudes of each charge and inversely proportional to the square of the distance between the charges. The fixed charge at time iteration (t) of particle i is shown as follows:

$$q_i(t) = \exp\left(-n * \left(f\left(x_i,t\right) - f\left(x_{best},t\right) / \left(\sum_{i=1}^{m}\left(f\left(x_i,t\right) - f\left(x_{best},t\right)\right)\right)\right)\right) \quad \forall i = 1,..,m \qquad (12)$$

Where t represents the iteration step, $q_i(t)$ is the charge of particle i at iteration t, $f(x_i,t)$, $f(x_{best},t)$, and $f(x_k,t)$ denote the objective value of particle i, the best solution, and particle k from m particles at time t; finally, n is the dimension of search space. The charge of each point i, $q_i(t)$, determines point's power of attraction or repulsion. Points (x_i) could be evaluated as a task into the graph representation (fig. 2).

The particles move along with total force and so diversified solutions are generated. The following formulation is the resultant force of particle i:

$$F_i(t) = \sum \begin{cases} \left(x_j(t) - x_i(t)\right) * \dfrac{\left(q_i(t) * q_j(t)\right)}{\left\|x_j(t) - x_i(t)\right\|^2} : f\left(x_j,t\right) < f\left(x_i,t\right) \\[4mm] \left(x_i(t) - x_j(t)\right) * \dfrac{\left(q_i(t) * q_j(t)\right)}{\left\|x_j(t) - x_i(t)\right\|^2} : f\left(x_j,t\right) \geq f\left(x_i,t\right) \end{cases}, \forall i = 1,...,m \qquad (13)$$

The following notes described an adapted version of EM for JSSP. According to this application, the initial population is obtained by choosing randomly from the list or pending tasks, as for the feasibility of solution, particles' path. The generic pseudo-code for the EM is reported in figure 6. Each particle is initially located into a source node (see disjunctive graph of figure 2). Particle is uniquely defined by a charge and a location into the node's space. Particle's position in each node is defined in a multigrid discrete set. While moving, particle jumps in a node based on its attraction force, defined in module and direction and way. If the force from starting line to arrival is in relation of positive inequality, the particles will be located in a plane position in linear dependence with force intensity. A selection mechanism could be set in order to decide where particle is directed, based on node force intensity. Force is therefore the resultant of particles acting in node. A solution for the JS is obtained only after a complete path from the source to the sink and the resulting force is updated according to the normalized makespan of different solutions.

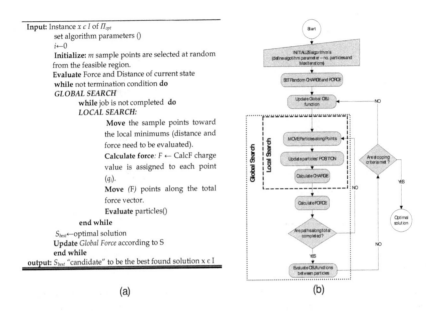

Figure 6. The Electromagnetic like Method; **6a**. the EM pseudo code; **6b**. the flow chart of a general EM procedure.

4.5. Simulated Annealing (SA)

The simulated annealing was presented by Scott Kirkpatrick *et al.* in 1983 [65] and by Vlado Černý in 1985 [66]. This optimization method is based on works of Metropolis *et al.*, [67] which allows describing the behaviour of a system in thermodynamic equilibrium at a certain temperature. It is a generic probabilistic metaheuristic used to find a good approximation to the global optimum of a given objective function. Mostly it is used with discrete problems such as the main part of the operations management problems.

Name and inspiration come from annealing in metallurgy, a technique that, through the heating and a controlled process of cooling, can increase the dimensions of the crystals inside the fuse piece and can reduce the defects inside the crystals structure. The technique deals with the minimization of the global energy E inside the material, using a control parameter called temperature, to evaluate the probability of accepting an uphill move inside the crystals structure. The procedure starts with an initial level of temperature T and a new random solution is generated at each iteration, if this solution improves the objective function, i.e., the E of the system is lower than the previous one. Another technique to evaluate the improvement of the system is to accept the new random solution with a likelihood according to a probability $exp(-\Delta E)$, where ΔE is the variation of the objective function. Afterwards a new iteration of the procedure is implemented.

As follows there is the pseudo-code of a general simulated annealing procedure:

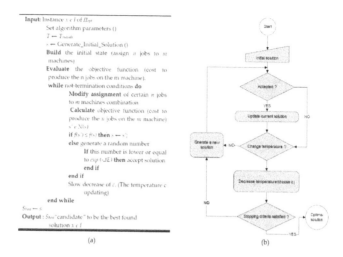

Input: Instance $x \in I$ of Π_{opt}
    Set algorithm parameters ()
    $T \leftarrow T_{initial}$
    $s \leftarrow$ Generate_Initial_Solution ()
    Build the initial state (assign n jobs to m machines)
    Evaluate the objective function (cost to produce the n jobs on the m machine).
    while not-termination conditions **do**
        Modify assignment of certain n jobs to m machines combination
        Calculate objective function (cost to produce the n jobs on the m machine)
        $s' \in N(s)$
        if $f(s') \leq f(s)$ **then** $s \leftarrow s'$;
        else generate a random number
            If this number is lower or equal to $exp(-\Delta E)$ **then** accept solution
            end if
        end if
        Slow decrease of t. (The temperature c updating)
    end while
$S_{best} \leftarrow s$
Output : S_{best} "candidate" to be the best found solution $x \in I$

(a)   (b)

Figure 7. The Simulated Annealing model; 7a. the SA pseudo code; 7b. the flow chart of a general SA procedure

For the scheduling issues, the application of the SA techniques requires the solutions fitness generated by each iteration, that is generally associated to the cost of a specific scheduling solution; the cost is represented by the temperature that is reduced for each iteration [68]. The acceptance probability can be measured as following:

$$Aij = \left\{ min \ \left[1, \ exp \left[\left(- \frac{C(j) - C(i)}{c} \right) \right] \right] \right\}$$
(14)

Another facet to be analysed is the stopping criteria, which can be fixed as the total number of iterations of the procedure to be computed.

4.6. Tabu Search (TS)

Tabu search (Glover, 1986) is an iterative search approach characterised by the use of a flexible memory [69]. The process with which tabu search overcomes local optimality is based on the evaluation function that chooses the highest evaluation solution at each iteration. The evaluation function selects the move, in the neighbourhood of the current solution, that produces the most improvement or the least deterioration in the objective function. Since, movement are accepted based on a probability function, a tabu list is employed to store characteristics of accepted moves so to classify them as taboo (i.e., to be avoided) in the later iteration. This is used to dodge cycling movements. A strategy called forbidding is employed to control and update the tabu list. This method was formalized by Glover [69]. An algorithm based on tabu search requires some elements: (i) the move, (ii) the neighbourhood, (iii) an initial solution, (iv) a search strategy, (v) a memory, (vi) an objective function and (vii) a stop criterion. The of TS is based on the definition of a first feasible solution S, which is stored as the current seed and the best solution, at each iteration, after the set of the neighbours is selected between the possible solutions deriving from the application of a movement. The value of the objective function is evaluated for all the possible movements, and the best one is chosen. The new solution is accepted even if its value is worse than the previous one, and the movement is recorded in a list, named taboo list.

For the problem of the scheduling in the job shops, generally a row of assignments of n jobs to m machines is randomly generated and the *cost* associated is calculated to define the fitness of the solution [70]. Some rules of movements can be defined as the crossover of some jobs to different machines and so on, defining new solutions and generating new values of the objective functions. The best solution between the new solutions is chosen and the movement is recorded in a specific file named taboo list. The stopping criterion can be defined in many ways, but simplest way is to define a maximum number of iterations [71].

In figure 8 are reported the pseudo-code and the flowchart for the application of TS to JSSP.

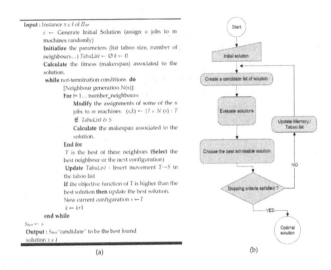

Figure 8. The Tabu Search approach; 8a. the TS pseudo code; 8b. the flow chart of a general TS procedure.

4.7. Neural Networks (NNs)

Neural networks are a technique based on models of biological brain structure. Artificial Neural Networks (NN), firstly developed by McCulloch and Pitts in 1943, are a mathematical model which wants to reproduce the learning process of human brain [72]. They are used to simulate and analyse complex systems starting from known input/output examples. An algorithm processes data through its interconnected network of processing units compared to neurons. Consider the Neural Network procedure to be a "black box". For any particular set of inputs (particular scheduling instance), the black box will give a set of outputs that are suggested actions to solve the problem, even though output cannot be generated by a known mathematical function. NNs are an adaptive system, constituted by several artificial neurons interconnected to form a complex network, those change their structure depending on internal or external information. In other words, this model is not programmed to solve a problem but it learns how to do that, by performing a *training* (or *learning) process* which uses a record of examples. This data record, called *training set*, is constituted by inputs with their corresponding outputs. This process reproduces almost exactly the behaviour of human brain that learns from previous experience.

The basic architecture of a neural network, starting from the taxonomy of the problems faceable with NNs, consists of three layers of neurons: the *input layer*, which receives the signal from the external environment and is constituted by a number of neurons equal to the number of input variables of the problem; the *hidden layer* (one or more depending on the complexity of the problem), which processes data coming from the input layer; and the *output layer*, which gives the results of the system and is constituted by as many neurons as the output variables of the system.

The error of NNs is set according to a testing phase (to confirm the actual predictive power of the network while adjusting the weights of links). After having built a training set of examples coming from historical data and having chosen the kind of architecture to use (among feed-forward networks, recurrent networks), the most important step of the implementation of NNs is the learning process. Through the training, the network can infer the relation between input and output defining the "strength" (weight) of connections between single neurons. This means that, from a very large number of extremely simple processing units (neurons), each of them performing a weighted sum of its inputs and then firing a binary signal if the total input exceeds a certain level (activation threshold), the network manages to perform extremely complex tasks. It is important to note that different categories of learning algorithms exists: (i) supervised learning, with which the network learns the connection between input and output thank to known examples coming from historical data; (ii) unsupervised learning, in which only input values are known and similar stimulations activate close neurons otherwise different stimulations activate distant neurons; and (iii) reinforcement learning, which is a retro-activated algorithm capable to define new values of the connection weights starting from the observation of the changes in the environment. Supervised learning by back error propagation (BEP) algorithm has become the most popular method of training NNs. Application of BEP in Neural Network for production scheduling is in: Dagli et al. (1991) [73], Cedimoglu (1993) [74], Sim et al. (1994) [75], Kim et al. (1995) [76].

The mostly NNs architectures used for JSSP are: searching network (Hopfield net) and *error correction network* (Multi Layer Perceptron). The Hopfield Network (a content addressable memory systems with weighted threshold nodes) dominates, however, neural network based scheduling systems [77]. They are the only structure that reaches any adequate result with benchmark problems [78]. It is also the best NN method for other machine scheduling problems [79]. In Storer *et al.* (1995) [80] this technique was combined with several iterated local search algorithms among which space genetic algorithms clearly outperform other implementations [81]. The technique's objective is to minimize the energy function E that corresponds to the makespan of the schedule. The values of the function are determined by the precedence and resource constraints which violation increases a penalty value. The Multi Layer Perceptron (i.e., MLP) consists in a black box of several layers allowing inputs to be added together, strengthened, stopped, non-linearized [82], and so on [83]. The black box has a great no. of knobs on the outside which can be filled with to adjust the output. For the given input problem, the training (network data set is used to adjust the weights on the neural network) is set as optimum target. Training an MLP is NP-complete in general.

In figure 9 it is possible to see the pseudo-code and the flow chart for the neural networks.

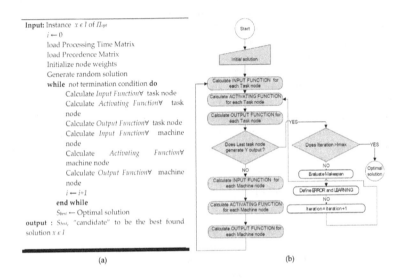

(a)   (b)

Figure 9. The NNs model; 9a. the implemented NNs pseudo code; 9b. the flow chart of generic NNs.

5. Discussion and conclusions

In this chapter, it was faced the most intricate problem (i.e., Job Shop) in order to explain approaches for scheduling in manufacturing. The JSP is one of the most formidable issues in the domain of optimization and operational research. Many methods were proposed, but only application of approximate methods (metaheuristics) allowed to efficiently solve large scheduling instances. Most of the best performing metaheuristics for JSSP were described and illustrated.

The likelihood of solving JSP can be greatly improved by finding an appropriate problem representation in computer domain. The acyclic graph representation is a quite good way to model alternatives in scheduling. How to fit approaches with problem domain (industrial manufacturing system) is generally a case in issue. Approaches are obviously affected by data and the results are subject to tuning of algorithm's parameters. A common rule is: less parameters generate more stable performances but local optimum solutions. Moreover, the problem has to be concisely encoded such that the job sequence will respect zoning and sequence constraints. All the proposed approaches use probabilistic transition rules and fitness information function of payoff (i.e., the objective function).

ACO and BE manifest common performances in JSSP. They do not need a coding system. This factor makes the approaches more reactive to the particular problem instance in issue. Notwithstanding, too many parameters have to be controlled in order to assure diversification of search. GAs surpasses their cousins in the request for robustness. The matching between genotype and phenotype across the schemata must be investigated in GAs in order to obtain promising results. The difficult of GA is to translate a correct phenotype from a starting genotype. A right balancing between crossover and mutation effect can control the perform- ance of this algorithm. The EM approach is generally affected by local stability that avoid global exploration and global performance. It is, moreover, subject to infeasibility in solutions because of its way to approach at the problem. SA and TS, as quite simpler approaches, dominate the panorama of metaheuristics proposal for JS scheduling. They manifest simplicity in imple- mentation and reduction in computation effort but suffer in local optimum falls. These approaches are generally used to improve performances of previous methodologies and they enhance their initial score. The influence of initial solutions on the results, for overall ap- proaches, is marked. Performances of NNs are generally affected by the learning process, over fitting. Too much data slow down the learning process without improving in optimal solution. Neural Network is, moreover, affected by difficulties in including job constraints with network representation. The activating signal needs to be subordinated to the constraints analysis.

Based on authors experience and reported paragraphs, it is difficult to definitively choose any of those techniques as outstanding in comparison with the others. Measurement of output and cost-justification (computational time and complexity) are vital to making good decision about which approach has to be implemented. They are vital for a good scheduling in operations management. In many cases there are not enough data to compare – benchmark instances, as from literature for scheduling could be useful - those methods thoroughly. In most cases it is evident that the efficiency of a given technique is problem dependent. It is possible that the parameters may be set in such way that the results of the algorithms are excellent for those benchmark problems but would be inferior for others. Thus, comparison of methods creates many problems and usually leads to the conclusion that there is no the only best technique. There is, however, a group of several methods that dominates, both in terms of quality of solutions and computational time. But this definition is case dependent.

What is important to notice here is: performance is usually not improved by algorithms for scheduling; it is improved by supporting the human scheduler and creating a direct (visual) link between scheduling actions and performances. It is reasonable to expect that humans will intervene in any schedule. Humans are smarter and more adaptable than computers. Even if users don't intervene, other external changes will happen that impact the schedule. Contingent maintenance plan and product quality may affect performance of scheduling. An algorithmic approach could be obviously helpful but it has to be used as a computerised support to the scheduling decision - evaluation of large amount of paths - where computational tractability is high. So it makes sense to see what optimal configuration is before committing to the final answer.

Author details

Marcello Fera[1], Fabio Fruggiero[2], Alfredo Lambiase[1], Giada Martino[1] and Maria Elena Nenni[3]

1 University of Salerno – Dpt. of Industrial Engineering, Fisciano (Salerno), Italy

2 University of Basilicata – School of Engineering, Potenza, Italy

3 University of Naples Federico II – Dpt. of Economic Management, Napoli, Italy

References

[1] Stutzle, T. G. Local Search Algorithms for Combinatorial Problems- Analysis, Algorithms and New Applications; (1998).

[2] Reeves, C. R. Heuristic search methods: A review. In D.Johnson and F.O'Brien Operational Research: Keynote Papers, Operational Research Society, Birmingham, UK, (1996). , 122-149.

[3] Trelea, I. C. The Particle Swarm Optimization Algorithm: convergence analysis and parameter selection.Information Processing Letters(2003). , 85(6), 317-325.

[4] Garey, M. R, & Johnson, D. S. Computers and intractability: a guide to the theory of NP-completeness, Freeman, (1979).

[5] Wight Oliver WProduction Inventory Management in the computer Age. Boston, Van Nostrand Reinhold Company, Inc., New York, (1974).

[6] Baker, K. R. Introduction to sequencing and scheduling, John Wiley, New York, (1974).

[7] Cox James F., John H. Blackstone, Jr., Michael S. Spencer editors,APICS Dictionary, American Production &Inventory Control Society, Falls Church, Virginia, (1992).

[8] Pinedo Michael, Scheduling Theory, Algorithms, and Systems, Prentice Hall, Englewood Cliffs, New Jersey, (1995).

[9] Hopp, W, & Spearman, M. L. Factory Physics. Foundations of manufacturing Management. Irwin/McGraw-Hill, Boston; (1996).

[10] Muth, J. F, & Thompson, G. L. Industrial Scheduling. Prentice-Hall, Englewood Cliffs, N.J., (1963).

[11] Garey, M. R, Johnson, D. S, & Sethi, R. The Complexity of Flow Shop and Job Shop Scheduling. Math. of Operation Research, (1976)., 2(2), 117-129.

[12] Blazewicz, J. Domschke W. and Pesch E..,The jobshops cheduling problem: conventional and new solution techniques,EJOR,1996; 93: 1-33.

[13] Aarts E.H.L., Van Laarhoven P.J.M., LenstraJ.K., and Ulder N.L.J. A computational study of local search algorithms for job shop scheduling. ORSA Journal on Computing, 1994; 6 (2): 118-125.

[14] Brucker, P. Scheduling Algorithms. Springer-Verlag, Berlin, (1995).

[15] Panwalkar, S. S. and Iskander Wafix. A survey of scheduling rules. Operations Research, (1977)., 25(1), 45-61.

[16] Carlier, J, & Pinson, E. An algorithm for solving the job-shop problem. Management Science, (1989)., 35(2), 164-176.

[17] Bertrand, J. W. M. The use of workload information to control job lateness in controlled and uncontrolled release production systems, J. of Oper. Manag., (1983). , 3(2), 79-92.

[18] GanttHenry L.. Work,Wages,andProfits,secondedition,Engineering Magazine Co., NewYork, (1916). Reprinted by Hive Publishing Company, Easton, Maryland, 1973.

[19] CoxJames F., John H. Blackstone, Jr., and Michael S. Spencer, edt.,APICS Dictionary, American Production &Inventory Control Society, Falls Church, Virginia, (1992).

[20] Roy, B, & Sussman, B. Les problèmes d'ordonnancement avec contraintes disjunctive, (1964).

[21] Fruggiero, F, Lovaglio, C, Miranda, S, & Riemma, S. From Ants Colony to Artificial Ants: A Nature Inspired Algorithm to Solve Job Shop Scheduling Problems. In Proc. ICRP-18; (2005).

[22] Adams, J, Balas, E, & Zawack, D. The shifting bottleneck procedure for job shop scheduling. Management Science, (1988). , 34

[23] Reeves, C. R. Modern Heuristic Techniques for Combinatorial Problems. John Wiley & Sons, Inc; (1993).

[24] Giffler, B. and Thompson G.L. Algorithms for solving productions cheduling problems. OperationsResearch, 1960;Vol.8: 487-503.

[25] Gere, W. S. Jr., Heuristics in Jobshop Scheduling, Manag. Science, (1966). , 13(1), 167-175.

[26] Rajendran, C, & Holthaus, O. A comparative Study of Dispatching rules in dynamics flowshops and job shops, European J. Of Operational Research. (1991). , 116(1), 156-170.

[27] Hertz, A, & Widmer, M. Guidelines for the use of meta-heuristics in combinatorial optimization, European Journal of Operational Research, (2003). , 151

[28] Zanakis, H. S, Evans, J. R, & Vazacopoulos, A. A. Heuristic methods and applications: a categorized survey, European Journal of Operational Research,(1989). , 43

[29] Gondran, M, & Minoux, M. Graphes et algorithmes, Eyrolles Publishers, Paris, (1985).

[30] Hubscher, R, & Glover, . Applying tabu search with influential diversification to multiprocessor scheduling, Computers Ops. Res. 1994; 21(8): 877-884.

[31] Blum, C, & Roli, A. Metaheuristics in combinatorial optimization: Overview and conceptual comparison,ACM Comput. Surv.,(2003). , 35, 268-308.

[32] Glover, F, & Kochenberger, G. A. Handbook of Metaheuristics, Springer, (2003).

[33] Kirkpatrick, S, Gelatt, C. D, & Vecchi, M. P. (1983). Optimization by Simulated Annealing. Science 2000; , 220(4598), 671-680.

[34] Birbil, S. I. Fang S An Electromagnetism-like Mechanism for Global Optimization. Journal of Global Optimization. (2003). , 25(3), 263-282.

[35] Mitchell, M. An Introduction to Genetic Algorithms. MIT Press, (1999).

[36] Glover, F, & Laguna, M. (1997). Tabu Search. Norwell, MA: Kluwer Academic Publ.

[37] Dorigo, M, & Di, G. Caro and L. M. Gambardella. Ant algorithm for discrete optimization. Artificial Life, (1999). , 5(2), 137-172.

[38] Pham, D. T, Ghanbarzadeh, A, Koc, E, Otri, S, Rahim, S, & Zaidi, M. The Bees Algorithm. Technical Note, Manufacturing Engineering Centre, CardiffUniversity, UK, (2005).

[39] Zhou, D. N, Cherkassky, V, Baldwin, T. R, & Olson, D. E. Aneuralnetwork approach to job-shop scheduling.IEEE Trans. on Neural Network,(1991).

[40] Holland John HAdaptation in Natural and Artificial Systems, University of Michigan,(1975).

[41] Darwin Charles "Origin of the species"(1859).

[42] Beck, J. C, Prosser, P, & Selensky, E. Vehicle Routing and Job Shop Scheduling: What's the difference?, Proc. of 13th Int. Conf. on Autom. Plan. and Sched. (ICAPS03); (2003).

[43] Bierwirth, C, Mattfeld, C, & Kopfer, H. On Permutation Representations for Scheduling Problems PPSN, (1996). , 310-318.

[44] Moon, I, & Lee, J. Genetic Algorithm Application to the Job Shop Scheduling Problem with alternative Routing, Industrial Engineering Pusan National Universit; (2000).

[45] Goss, S, Aron, S, Deneubourg, J. L, & Pasteels, J. M. Self-organized shortcuts in the Argentine ant. Naturwissenschaften. (1989)., 76, 579-581.

[46] Van Der Zwaan, S, & Marques, C. Ant colony optimization for job shop scheduling. In Proc. of the 3rd Workshop on genetic algorithms and Artificial Life (GAAL'99), (1999).

[47] Dorigo, M, Maniezzo, V, & Colorni, A. The Ant System: Optimization by a colony of cooperating agents. IEEE Transactions on Systems, (1996)., 26, 1-13.

[48] Cornea, D, Dorigo, M, & Glover, F. editors. New ideas in Optimization. McGraw-Hill International, Published in (1999).

[49] Colorni, A, Dorigo, M, Maniezzo, V, & Trubian, M. Ant system for Job-shop Scheduling". JORBEL-Belgian Journal of Operations Research, Statistics and Computer Science, (1994).

[50] Gould, J. L. Honey bee recruitment: the dance-language controversy.Science.(1975).

[51] Grosan, C, Ajith, A, & Ramos, V. Stigmergic Optimization: Inspiration, Technologies and Perspectives.Studies in Computational Intelligence.(2006). Springer Berlin/ Heidelberg., 31

[52] Camazine, S, Deneubourg, J, Franks, N. R, Sneyd, J, Theraula, G, & Bonabeau, E. Self-Organization in Biological Systems. Princeton: Princeton University Press, (2003).

[53] Von Frisch, K. Bees: Their Vision, Chemical Senses and Language. (Revised edn) Cornell University Press, N.Y., Ithaca, (1976).

[54] Riley, J. R, Greggers, U, Smith, A. D, Reynolds, D. R, & Menzel, R. The flight paths of honeybees recruited by the waggle dance". Nature(2005)., 435, 205-207.

[55] Eberhart, R, & Shi, . ., SwarmIntelligence.Morgan Kaufmann, San Francisco, 2001.

[56] Seeley, T. D. The wisdom of the Hive: The Socal Physiology of Honey Bee Colonies. Massachusetts: Harward University Press, Cambridge, (1996).

[57] Tuba, M. Artificial BeeColony(ABC) with crossover and mutation. Advances in computer Science, (2012)., 157-163.

[58] Chong CS Low, MYH Sivakumar AI, Gay KL. Using a Bee Colony Algorithm for Neighborhood Search in Job Shop Scheduling Problems, In 21st European Conference On Modelling and Simulation ECMS (2007).

[59] Karaboga, D, & Basturk, B. On The performance of Artificial Bee Colony (ABC) algorithm. Applied Soft Computing, (2008).

[60] Pham, D. T, Ghanbarzadeh, A, Koc, E, Otri, S, Rahim, S, & Zaidi, M. The Bees Algo-
      rithm- A Novel Tool for Complex Optimisation Problems, Proceedings of IPROMS
      (2006). Conference, , 454-461.

[61] Birbil, S. I. Fang S An Electromagnetism-like Mechanism for Global Optimization.
      Journal of Global Optimization. (2003). , 25(3), 263-282.

[62] Coulomb Premier mémoire sur l'électricité et le magnétismeHistoire de l'Académie
      Royale des Sciences, , 569-577.

[63] Durney Carl H. and Johnson, Curtis C. Introduction to modern electromagnetics.
      McGraw-Hill. (1969).

[64] Griffiths David J.Introduction to Electrodynamics(3rd ed.). Prentice Hall; (1998).

[65] Kirkpatrick, S, Gelatt, C. D, & Vecchi, M. P. Optimization by Simulated Annealing.
      Science. (1983). , 220(4598), 671-680.

[66] Cerný, V. Thermodynamical approach to the traveling salesman problem: An effi-
      cient simulation algorithm. J. of Optimization Theory and Applications. (1985). , 45,
      41-51.

[67] Metropolis Nicholas; Rosenbluth, Arianna W.; Rosenbluth, Marshall N.; Teller, Au-
      gusta H.; Teller, Edward. Equation of State Calculations by Fast Computing Ma-
      chines. The Journal of Chemical Physics. (1953).

[68] Van Laarhoven P.J.M, Aarts E.H.L., and J. K. Lenstra. Job shop scheduling by simula-
      tedannealing. Operations Research, Vol. 40, No. 1, pp. 113-125, 1992.

[69] Glover, F. Future paths for Integer Programming and Links to Artificial Intelligence.
      Computers and Operations Research (1986). , 5(5), 533-549.

[70] Dell'Amico M and Trubian M. Applying tabu search to the job-shop scheduling
      problem. Annals of Operations Research, (1993). , 41, 231-252.

[71] Taillard, E.D. Parallel taboo search techniques for the job-shop scheduling problem.
      ORSAJournal on Computing, 1994; 6(2): 108-117.

[72] Marquez, L, Hill, T, Connor, O, & Remus, M. W., Neural network models for forecast
      a review. In: IEEE Proc of 25th Hawaii International Conference on System Sciences.
      (1992). , 4, 494-498.

[73] Dagli, C. H, Lammers, S, & Vellanki, M. Intelligent scheduling in manufacturing us-
      ing neural networks, Journal of Neural Network Computing Technology Design and
      Applications, (1991). , 2(4), 4-10.

[74] Cedimoglu, I. H. Neural networks in shop floor scheduling, Ph.D. Thesis, School of
      Industrial and Manufacturing Science, Cranfield University, UK. (1993).

[75] Sim, S. K, Yeo, K. T, & Lee, W. H. An expert neural network system for dynamic job-shop scheduling, International Journal of Production Research, (1994). , 32(8), 1759-1773.

[76] Kim, S. Y, Lee, Y. H, & Agnihotri, D. A hybrid approach for sequencing jobs using heuristic rules and neural networks, Prod. Planning and Control, (1995). , 6(5), 445-454.

[77] Hopfield, J. J, & Tank, D. W. Neural computational of decisions in optimization problems. Biological Cybernetics, (1985). , 52, 141-52.

[78] Foo, S. Y, & Takefuji, Y. Stochastic neural networks for solving job-shop scheduling: Part 1. Problem representation, in: Kosko B, IEEE International Conference on Neural Networks, San Diego, CA, USA, (1988). , 1988, 275-282.

[79] Haykin S Neural networks: a comprehensive foundation nd edn. Prentice Hall, New Jersey; (2001).

[80] Storer, R. H, Wu, S. D, & Vaccari, R. Problem and heuristic space search strategies for job shop scheduling, ORSA Journal on Computing, (1995). , 7(4), 453-467.

[81] Van Hulle, M. M. A goal programming network for mixed integer linear programming: A case study for the jobshop scheduling problem, International Journal of Neural Systems, (1991).

[82] Leshno, M, Lin, V. Y, Pinkus, A, & Schocken, S. Multilayer feedforward networks with a non-polynomial activation function can approximate any function. Neural Net. (1993). , 6(6), 861-867.

[83] Karlik, B, & Olgac, A. V. Performance analysis of various activation functions in generalized MLP architectures of neural networks. Int J. Artif. Int. Expert Syst. (2011). , 1(4), 111-122.

Chapter 6

On Just-In-Time Production Leveling

Francesco Giordano and Massimiliano M. Schiraldi

Additional information is available at the end of the chapter

1. Introduction

Since the 80's, the Japanese production techniques and philosophies spread among the Western manufacturing companies. This was possible because the Toyota Motor Company experience was indeed a success. The so-called "Toyota Production System" (TPS) seemed to be the "one best way" to manage a manufacturing production site.

On the other side, it is also well known that not every implementation of Lean Production was a success, especially in Western companies: some enterprises – together with the consultancy firms that should have supported them – forgot that there are some main hypotheses and issues to comply with, in order to achieve Toyota-like results. On top of this, certain requisites are not related to a mere managerial approach, but depend on exogenous conditions, e.g. market behavior or supplier location; thus, not every company can successfully implement a TPS system.

One critical requirement for a TPS approach to be effective is that the production plan should be leveled both in quantity and in mix. This is indicated by the Japanese term *heijunka* (平準化), which stands for "leveling" or "smoothing". Here, we will focus our attention on why leveled production is a key factor for JIT implementation, and specifically we will describe and analyze some approaches to deal with the leveling problem.

At first, the original Toyota Production System is briefly recalled, with specific regard to the *Just In Time* (JIT) approach to manage inventories in production. JIT is a stock replenishment policy that aims to reduce final product stocks and work-in-process (WIP); it coordinates requirements and replenishments in order to minimize stock-buffer needs, and it has reversed the old make-to-stock production approach, leading most companies to adopt "pull" instead of "push" policies to manage material and finished product flows. However, in case of unleveled demand, stock levels in JIT may grow uncontrolled.

Secondly, *kanban*-based production is described: *kanban*, a Japanese word meaning "visual record", is a card that contains information on a product in a given stage of the manufacturing process, and details on its path of completion. It is acknowledged as one of the most famous technique for material management in the JIT approach. Here we will present some common algorithms for managing *kanban* queues, along with their criticalities in terms of production smoothing requirements and reduced demand stochasticity. Some of the JIT-derivative approaches will be recalled as well: CONWIP, Quick Response Manufacturing, Theory of Constraints and the Just-In-Sequence approach.

Then, a review on the mixed-model JIT scheduling problem (MMJIT), along with the related solving approaches, is presented. Despite the huge literature on MMJIT mathematical programming approaches, here it will be described why the real-world production systems still prefer the simpler kanban approach and the old (1983) Goal Chasing Method algorithm. In the end, an overview on simulators advantages to test alternative heuristics to manage JIT production is presented.

2. Managing Just-In-Time production systems

Just-in-Time was first proposed within the *Toyota Production System* (TPS) by Taiichi Ohno after the 50's when he conceived a more convenient way to manage inventory and control production systems [1]. *Lean Production* – the un-branded name of TPS – is a mix of a philosophy for production systems management and a collection of tools to improve the enterprise performances [2]. Its cornerstones are the reduction of *muda* (wastes), *mura* (unevenness) and *muri* (overburden). Ohno identified seven wastes [3] that should be reduced to maximize the return of investment of a production site:

* transportation;

* inventory;

* motion;

* waiting;

* over-processing;

* over-producing;

* defects.

The TPS catchphrase emphasizes the "zero" concept: zero machine changeovers ("set-ups"), zero defects in the finished products, zero inventories, zero production stops, zero bureaucracy, zero misalignments. This result may be reached through a continuous improvement activity, which takes cue from Deming's *Plan-Do-Check-Act* cycle [1]: the *kaizen* approach.

Just-In-Time is the TPS solution to reduce inventory and waiting times. Its name, according to [4], was coined by Toyota managers to indicate a method aimed to ensure "the right products, in the right quantities, *just in time*, where they are needed". Differently from Orlicky's Material Requirement Planning (MRP) – which schedules the production run in advance compared to the moment in which a product is required [5] – JIT approach will replenish a stock only after its depletion. Among its pillars there are:

- one-piece flow;

- mixed-model production;

- demand-pull production;

- *takt* time;

Indeed, generally speaking, processing a 10 product-batch requires one tenth of the time needed for a 100 product-batch. Thus, reducing the batch value (up to "one piece") would generate benefits in reducing either time-to-market or inventory level. This rule must come along with mixed-model production, which is the ability of manufacture different products alternating very small batches on shared resources. Demand-pull production indicates that the system is activated only after an order receipt; thus, no semi-finished product is processed if no downstream workstation asks for it. On top of this, in order to smooth out the material flow, the process operations should be organized to let each workstation complete different jobs in similar cycle times. The base reference is, thus, the *takt* time, a term derived from the German word *taktzeit* (cycle time), which is computed as a rapport between the net operating time, available for production, and the demand in terms of units required. These are the main differences between the *look-ahead* MRP and the *look-back* JIT system. For example, the MRP algorithm includes a lot-sizing phase, which results in product batching; this tends to generate higher stock levels compared to the JIT approach. Several studies have been carried out on MRP lot-sizing [6] and trying to improve the algorithm performance [7, 8, 9]; however, it seems that JIT can outperform MRP given the *heijunka* condition, in case of leveled production both in quantity and in mix. The traditional JIT technique to manage production flow is named *kanban*.

3. The kanban technique

A *kanban* system is a multistage production scheduling and inventory control system [10]. Kanban cards are used to control production flow and inventories, keeping a reduced production lead time and work-in-process. Clearly, a kanban is not necessarily a physical paper/plastic card, as it can be either electronic or represented by the container itself.

Since it was conceived as an easy and cheap way to control inventory levels, many different implementations of kanban systems have been experimented in manufacturing companies all over the world. In the following paragraphs, the most commonly used "one/two cards" kanban systems are described.

3.1. One-card kanban system

The "one-card" is the simplest implementation of kanban systems. This approach is used when the upstream and downstream workstations (respectively, the preceding and succeeding processes) are physically close to each other, so they can share the same stock buffer. The card is called "Production Order Kanban" (POK) [11, 12]. The stock buffer acts either as the outbound buffer for the first (A) workstation or as the inbound buffer for the second (B) workstation. A schematic diagram of a one-card system is shown in Figure 1.

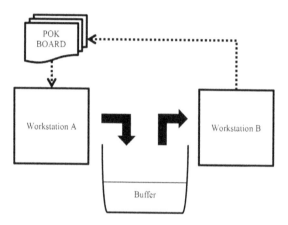

Figure 1. A one-card kanban system

Here, each container (the JIT unit load) has a POK attached, indicating the quantity of a certain material contained, along with eventual complementary information. The POK also represents a production order for the Workstation A, indicating to replenish the container with the same quantity. When a B operator withdraws a container from the buffer, he removes the POK from the container and posts it on a board. Hence, A operator knows that one container with a specific part-number must be replenished in the stock buffer.

3.2. Two-card kanban system

In the two-card system, each workstation has separate inbound and outbound buffers [13, 14]. Two different types of cards are used: Production Order Kanbans (POK) and Withdrawal Kanbans (WK). A WK contains information on how much material (raw materials / semi-finished materials) the succeeding process should withdraw. A schematic diagram of a two-card system is shown in Figure 2.

Each work-in-progress (WIP) container in the inbound buffer has a WK attached, as well as each WIP in the outbound buffer has a POK. WK and POK are paired, i.e. each given part number is always reported both in n POK and n WK. When a container is withdrawn from the inbound buffer, the B operator posts the WK on the WK board. Then, a warehouse-keeper operator uses the WK board as a picking list to replenish the inbound buffer: he takes the WK off the board and look for the paired POK in the outbound buffer. Then, he moves the corresponding quantity of the indicated material from the A outbound to the B inbound buffer, while exchanging the related POK with the WK on the container, restoring the initial situation. Finally, he posts the left POK on the POK board. Hence, like in the previous scenario, A workstation operator knows that one container of that kind must be replenished in the outbound stock buffer. The effectiveness of this simple technique – which was described in details by several authors [3, 14, 15, 16] – is significantly influenced by the policy followed to determine the kanban processing order, in the boards.

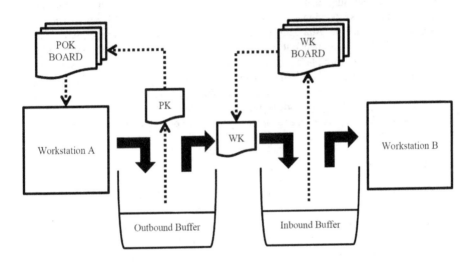

Figure 2. A two-card kanban system

3.3. Standard approaches to manage the kanban board

From the previously described procedure, it is clear that the each workstation bases its production sequence on kanban cards posted on the POK board. In literature, few traditional ways to manage the board are reported: each of them is quite easy to implement and does not require significant investments in technology or other expensive assets.

The most commonly used policy [3] requires having a board for each station, and this should be managed as a single First-In-First-Out (FIFO) queue. The board is usually structured as one vector (one column, multiple rows): POK are posted on the board in the last row. Rows are grouped in three zones (red/yellow/green) which indicate three levels of urgency (respectively, high/medium/low). Kanban are progressively moved from the green to the red zone and the workstation operator will process the topmost kanban. If a kanban reaches the red rows, it means that the correspondent material is likely to be requested soon, by the succeeding process. Thus, it should be urgently replenished in the outbound buffer, in order to avoid stock-outs.

Although this policy does not rely on any optimized procedure, it may ensure a leveled production rate in each workstation, given the fact that other TPS pillars are implemented, e.g. setup time reduction and mixed model scheduling. Indeed, if the final downstream demand is leveled, the production plan of the workstations will be leveled as well. Clearly, this policy is vulnerable to high setup times and differences among workstations cycle times: in this latter case, indeed, the ideal jobs sequence for a workstation may be far from optimal for the preceding. It is noticeable that the colored zones on the board only provide a visual support for the operators and do not influence the jobs processing order.

A *heijunka box* is a sort of enhanced kanban board: it still acts as a visual scheduling tool to obtain production leveling at the workstations. However, differently from the traditional board, it manages to keep evidence of materials distinctions. Usually, it is represented as a grid-shaped wall schedule. Analogously to the simpler board, each row represents a time interval (usually, 30-60 minutes), but multiple columns are present, each one associated to a different material. POKs are placed in the so-called "pigeon-holes" within the box, based on number of items to be processed in the job and on the material type. Workstation operators will process all the kanban placed in the current period row, removing them from the box. Hence, heijunka box not only provides a representation for each job queued for production, but for its scheduled time as well, and allows operators to pursue production leveling when inserting new POKs in the boxes.

3.4. Criticism on JIT

During the last decades, Just-In-Time has been criticized from different authors [17]. Indeed, certain specific conditions – which, though, are not uncommon in manufacturing companies – can put in evidence some well-known weak points of the Japanese approach. Specifically, un-steady demand in multi-product environments where differences in processing lead times are not negligible represent a scenario where JIT would miserably fail, despite the commitment of the operations managers.

First, we have to keep in mind that one pillar of Lean Production is the "one-piece-flow" diktat. A one-piece batch would comply with the Economic Production Quantity theory [18] only when order cost (i.e. setup time) is zero. Having non-negligible setup times hampers JIT implementation and makes the production leveling problem even more complicated. It is peculiar that, originally, operations researchers concentrated on finding the best jobs sequence considering negligible setups time. This bound was introduced into the mixed model kanban scheduling problem only since 2000. Setups are inevitable in the Lean Production philosophy, but are considered already optimized as well. Given that setup times are *muda*, TPS approach focuses on quickening the setup time, e.g. through technical interventions on workstations or on the setup process with SMED techniques, not on reducing their frequency: the increased performance gained through setups frequency reduction is not worth the flexibility loss that the system may suffer as a consequence. Indeed, the standard kanban management system, ignoring the job sequencing, does not aim at reducing setup wastes at all. Analogously, the Heijunka box was developed for leveling production and can only assure that the product mix in the very short term reproduces that in the long term; in its original application, the decision on the job sequence is left to the operator. Only in some enhanced version, the sequence is predefined applying some scheduling algorithm.

Given the fact that JIT is based on stock replenishment, constant production and withdrawal rates should be ensured in order to avoid either stock outs or stock proliferation. Mixed-model production requires a leveled Master Production Schedule (MPS) [19], but this is not sufficient to smooth the production rate in a short time period. While it is easy to obtain a leveled production in a medium or even medium-short period, it is difficult to do it in each hour, for each workstation and each material.

Indeed, demand is typically unstable under two points of view: random frequency, which is the chance that production orders are irregularly received, and random quantities, which is related to product mix changes. Indeed, since TPS assume minimal stock levels, the only chance to cope with demand peak is to recur to extra production capacity. However, available production capacity should be higher than required as the average (as TPS requires), but for sure cannot be limitless. Thus, the JIT management system should anyway be able to consider the opportunity of varying the maintenance plan as well as the setup scheduling, in case of need. On the other hand, if the production site faces a leveled production, changes in product mix should not represent a problem; however, they increase sequencing problem complexity. Most of the operational research solutions for JIT scheduling are designed for a fixed product mix, thus its changes can greatly affect the optimality of solutions, up to make them useless.

On the contrary, kanban board mechanism is not influenced by demand randomness: as long as demand variations are contained into a certain (small) interval, kanban-managed workstations will handle their production almost without any problem. Therefore, in case of unstable demand, in order to prevent stock-outs, inventory managers can only increase the kanban number for each product: the greater are the variations, the greater is the need of kanban cards and, thus, the higher is the stock level. In order to prevent stock level raise, some authors [20, 21] proposed to adopt a frozen schedule to implement JIT production in real companies, where demand may clearly be unstable. Anyway, this solution goes in the opposite direction compared to JIT foundations.

Moreover, one-piece-flow conflicts with demand variability: the batch size should be chosen as its processing time exceeds the inter-arrival time of materials requests. Thus, the leveling algorithm must find the proper sequencing policy that, at the same time, reduces the batch size and minimize the inter-arrival time of each material request. This sequence clearly depends on the total demand of each material in the planning horizon. However, JIT does not use forecasting, except during system design; thus, scheduling may be refreshed daily. From a computational point of view, this is a non-linear integer optimization problem (defined *mixed-model just-in-time scheduling problem*, MMJIT), which has non-polynomial complexity and it currently cannot be solved in an acceptable time. Thus, reliable suppliers and a clockwork supply chain are absolutely required to implement JIT. Toyota faced this issue using various approaches [22]:

- moving suppliers in the areas around the production sites, in order to minimize the supply lead time;

- collaborating with the suppliers and helping them to introduce JIT in their factories;
- always relying on two alternative suppliers for the same material, not to be put in a critical situation.

In the end, it should be noted that, considering that at each stage of the production process at least one unit of each material must be in stock, in case of a great product variety the total stock amount could be huge in JIT. This problem was known also by Toyota [1], who addressed it limiting the product customization opportunities and bundling optional combinations.

4. Alternative approaches

4.1. CONWIP

Many alternatives to JIT have been proposed since TPS appeared in Western countries. One of the most famous JIT-derivative approaches is CONWIP (CONstant Work-In-Process). This methodology, firstly proposed in the 90's [23], tries to mix push and pull approaches: it schedules tasks for each station – with a push approach – while production is triggered by inventory events, which is a pull rule. Thus, CONWIP is card-based, as kanban systems, but cards do not trigger the production of a single component in the closest upward workstation; conversely, cards are used to start the whole production line, from beginning downwards. Then, from the first workstation up to the last one, the process is push-driven; materials are processed as they get to an inbound buffer, notwithstanding the stock levels. Only the last workstation has a predetermined stock level, similar to the JIT outbound buffer. All queues are managed through a FIFO policy. In order to have a leveled production rate and to avoid production spikes or idle times, the system is calibrated on the slowest workstation, the *bottleneck*. Results from simulations showed [24] that CONWIP could grant shorter lead times and more stable production rate if compared to Kanban; however, it usually needs a higher WIP level. A CONWIP system is also easier to implement and adjust, since it has only one card set.

4.2. POLCA

Another alternative technique mixing push and pull system is the POLCA (Paired-Cell Overlapping Loops of Cards with Authorization), which stands at the base of the Quick Response Manufacturing (QRM) approach, proposed in 1998 [25]. QRM aims to minimize lead times rather than addressing waste reduction, as TPS does. A series of tools, such as manufacturing critical-path time, cellular organization, batch optimization and high level MRP, are used to minimize stock levels: the lesser is the lead time, the lesser is the on-hand inventory. Likewise CONWIP, POLCA handles WIP proliferation originating from multiple products, since it does not require each station to have a base stock of each component. At first, an MRP-like algorithm (called HL/MRP) creates some "Release Authorization Times". That means that the HL/MRP system defines when each cell may start each job, as MRP defines the "Start Dates". However, differently from a standard push system - where a workstation should process the job as soon as possible - POLCA simply authorizes the possibility to start the job. Analogously to CONWIP and Kanban, POLCA uses production control cards in order to control material flows. These cards are only used between, and not within, work cells. Inside each work cell, material flows resemble the CONWIP approach. On top of this, the POLCA cards, instead of being specifically assigned to a product as in a Kanban system, are assigned to pairs of cells. Moreover, whereas a POK card is an inventory replenishment signal, a POLCA card is a capacity signal. If a card returns from a downstream cell, it signals that there is enough capacity to process a job. Thus, the preceding cell will proceed only if the succeeding cell has available production capacity. According to some authors [20] a POLCA system may overcome the drawbacks of both standard MRPs and kanban systems, helping in managing both short-term fluctuation in capacity (slowdowns, failures, setups, quality issues) and reducing unnecessary stocks, which is always present in any unlevelled replenishment system – i.e. where heijunka condition is not met.

4.3. Just in sequence

The Just in Sequence approach is an evolution of JIT, which embeds the CONWIP idea of mixing push/requirement and pull/replenishment production management systems. The overall goal of JIS is to synchronize the material flow within the supply chain and to reduce both safety stocks and material handling. Once the optimal production sequence is decided, it is adopted all along the process line and up to the supply chain. Thus, the suppliers are asked to comply not only to quantity requirements but also to the product sequence and mix, for a certain period of time. In this case the demand must be stable, or a frozen period should be defined (i.e. a time interval, prior to production, in which the demand cannot be changed) [26]. Clearly, when the demand mix significantly changes, the sequence must be re-computed, similarly to what happens in MRP. This makes the JIS system less flexible compared to JIT. Research results [27] proved that, applying some techniques to reduce unsteadiness – such as flexible order assignment or mixed bank buffers – the sequence can be preserved with a low stock level. Thanks to *ad-hoc* rescheduling points the sequence can be propagated downstream, reducing the impact of variability.

4.4. The "Theory of Constraints" approach

Leveraging on the common idea that "a chain is no stronger than its weakest link", the Israeli physicist E.M. Goldratt firstly introduced the Theory of Constraints (TOC) in his most famous business novel "the Goal" [28]. Looking to a production flow-shop as a chain, the weakest link is represented by the line bottleneck. Compared to the TPS approach of reducing wastes, this approach is focused on improving bottleneck operations, trying to maximize the *throughput* (production rate), minimizing inventory and operational expenses at the same time.

Its implementation is based on a loop of five steps:

1. constraint identification;

2. constraint optimization;

3. alignment of the other operations to the constraint optimization;

4. elevation of the constraint (improving throughput);

5. if the constraint after the previous 4 steps has moved, restart the process.

Again, Deming's concept of "improvement cycle" is recalled. However, improvements are only focused on the bottleneck, the Critical Constraint Resource (CCR), whereas in the Lean Production's Kaizen approach bottom-up, an improvement may arise wherever wastes are identified; moreover, improvements only aim to increase throughput. It is though noticeable that the author includes, as a possible throughput constraint, not only machinery problem but also people (lack of proper skills) and policies (bad working). To this extent, Goldratt coined the "Drum-Buffer-Rope" (DBR) expression: the bottleneck workstation will define the production takt-time, giving the beat as with a drum. The remaining upstream and down-stream workstations will follow this beat. This requires the drum to have an optimized schedule, which is imposed to all the production line. Thus, takt-time is not defined from the final demand anymore, but is set equal to the CCR minimal cycle time, given that the bottleneck capacity cannot be overcome. A "buffer" stock is only placed before the CCR, assuring that no upward issue could affect the process pace, reducing line throughput. This helps in reducing the inventory level in comparison to replenishment approaches, where buffers are placed among all the workstations. Eventually, other stock buffers may be placed in few synchronization points in the processes, besides the final product warehouse, which prevents stock-outs due to oscillating demand. The "rope" represents the job release authorization mechanism: a CONWIP approach is used between the CCR and the first phase of the process. Thus, the advanced entrance of a job in the system is proportional to the buffer size, measured in time. Failing to comply with this rule is likely to generate too high work-in-process, slowing down the entire system, or to generate a starvation condition on the CCR, with the risk of reducing the throughput. Several authors [29, 30, 31] analyzed the DBR rule in comparison to planning with mathematical linear programming techniques. Results on the most effective approach are controversial.

5. The mixed-model JIT scheduling problem

The leveling problem in JIT operations research literature was formalized in 1983 as the "mixed-model just-in-time scheduling (or sequencing) problem" (MMJIT) [32], along with its first solution approach, the "Goal Chasing Method" (GCM I) heuristic.

Some assumptions are usually made to approach this problem [33]. The most common are:

- no variability; the problem is defined in a deterministic scenario;

- no details on the process phases: the process is considered as a black box, which transforms raw materials in finished products;

- zero setup times (or setup times are negligible);

- demand is constant and known;

- production lead time is the same for each product.

Unfortunately, the problem with these assumptions virtually never occurs in industry. However, the problem is of mathematical interest because of its high complexity (in a theoretical mathematical sense). Because researchers drew their inspiration from the literature and not from industry, on MMJIT far more was published than practiced.

The objective of a MMJIT is to obtain a leveled production. This aim is formalized in the Output Rate Variation (ORV) objective function (OF) [34, 35]. Consider a M set of m product models, each one with a d_m demand to be produced during a specific period (e.g., 1 day or shift) divided into T production cycles, with

$$\sum_{m \in M} d_m = T$$

Each product type m consists of different components p belonging to the set P. The production coefficients a_{pm} specify the number of units of part p needed in the assembly of one unit of product m. The matrix of coefficients A = (a_{pm}) represents the Bill Of Material (BOM). Given the total demand for part p required for the production of all m models in the planning horizon, the target demand rate r_p per production cycle is calculated as follows:

$$r_p = \frac{\sum_{m \in M} d_m \cdot a_{pm}}{T}, \ \forall \, p \in P$$

Given a set of binary variables x_{mt} which represent whether a product m will be produced in the t cycle, the problem is modeled as follows [33]:

$$\min Z = \sum_{p \in P} \sum_{t=1}^{T} \left(\sum_{m \in M} \sum_{t=1}^{t} x_{mt} \cdot a_{pm} - t \cdot r_p \right)^2$$

subject to

$$\sum_{m \in M} x_{mt} = 1, \ \forall \, t = 1, ..., T$$

$$\sum_{t=1}^{T} x_{mt} = d_m, \ \forall \, m \in M$$

$$x_{mt} \in \{0, 1\}, \ \forall \, m \in M; t = 1, ..., T$$

The first and second group of constraints indicate that for each time t exactly one model will be produced and that the total demand d_m for each model will be fulfilled by the time T. More constraints can be added if required, for instance in case of limited storage space.

A simplified version of this problem, labeled "Product Rate Variation Problem" (PRV) was studied by several authors [36, 37, 38], although it was found it is not sufficient to cope with the variety of production models of modern assembly lines [33]. Other adaptations of this problem were proposed along the years; after 2000, when some effective solving algorithms were proposed [39], the literature interest moved on to the MMJIT scheduling problem *with setups* [40]. In this case, a dual OF is used [41]: the first part is the ORV/PRV standard function, while the second is simply:

$$\min S = 1 + \sum_{t=2}^{T} s_t$$

In this equation, $s_t = 1$ if a setup is required in position t; while $s_{t-}\,0$, if no setup is required. The assumptions of this model are:

- an initial setup is required regardless of sequence; this is the reason for the initial "1" and the t index follows on "2";

- the setup time is standard and it is not dependent from the product type;

- the setup number and setup time are directly proportional each other.

The following sets of bounds must be added in order to shift s_t from "0" to "1" if the production switches from a product to another:

$$x_{m(t-1)} - x_{mt} \le s_t, \quad \forall\, t = 2, \dots, T, \ \forall\, m \in M$$

$$s_t \in \{0, 1\}, \quad \forall\, t = 2, \dots, T$$

Being a multi-objective problem, the MMJIT with setups has been approached in different ways, but it seems that no one succeeded in solving the problem using a standard mathematical approach. A simulation approach was used in [42]. Most of the existing studies in the literature use mathematical representations, Markov chains or simulation approaches. Some authors [10, 40] reported that the following parameters may vary within the research carried out in the recent years, as shown in Table 1 below.

5.1. A review on solving approaches

The MMJIT problem, showing nonlinear OF and binary variables, has no polynomial solutions as far as we know. However, a heuristic solution approach can be effective. To get to a good solution, one among dynamic programming, integer programming, linear programming, mixed integer programming or nonlinear integer programming (NLP) techniques can be used. However, those methodologies usually require a long time to find a solution, so are infrequently used in real production systems [44]. Just a few studies used other methods such as statistical analysis or the Toyota formula [45]. The most renowned heuristics are the Miltenburg's [36] and the cited Goal Chasing Method (GCM I) developed in Toyota by Y. Monden. Given the products quantities to be processed and the associated processing times, GCM I computes an "average consumption rate" for the workstation. Then, the processing sequence is defined choosing each successive product according to its processing time, so that the cumulated consumption rate "chases" its average value. A detailed description of the algorithm can be found in [32]. GCM I was subsequently refined by its own author, resulting in the GCM II and the Goal Coordination Method heuristics [46].

The most known meta-heuristics to solve the MMJIT [44, 47, 48] are:

- Simulated Annealing;

- Tabu Search;

- Genetic Algorithms;

- Scalar methods;

- Interactive methods;

- Fuzzy methods;

- Decision aids methods;

- Dynamic Programming.

Parameter / major alternatives		Alternatives				
Model structure		Mathematical programming	Simulation	Markov Chains	Other	
Decision variables		Kanban number	Order interval	Safety Stock level	Other	
Performance measures		Kanban number	Utilization ratio		Leveling effectiveness	
Objective function	Minimize cost	Setup cost	Inventory holding cost	Operating cost	Stock-out cost	
	Minimize inventory					
	Maximize throughput					
Setting	Layout	Flow-shop	Job-shop		Assembly tree	
	Period number	Multi-period		Single-period		
	Item number	Multi-item		Single-item		
	Stage number	Multi-stage		Single-stage		
	Machine number	Multiple machines		Single machine		
	Resources capacity	Capacitated		Non-capacitated		
Kanban type		One-card		Two-card		
Assumptions	Container size	Defined		Ignored (container size equals one item)		
	Stochasticity	Random set-up times	Random demand	Random lead times	Random processing times	Determinism
	Production cycles	Manufacturing system		Continuous production		
	Material handling	Zero withdrawal times		Non-zero withdrawal times		
	Shortages	Ignored		Computed as lost sales [43]		
	System reliability	Dynamic demand	Breakdowns possibility	Imbalance between stages	Reworks	Scraps

Table 1. Alternative configurations of most common MMJIT models

In some experiments [44] Tabu Search and Simulated Annealing resulted to be more effective than GCM; however, the computational complexity of these meta-heuristics – and the consequent slowness of execution – makes them quite useless in practical cases, as the same authors admitted.

Another meta-heuristic based on an optimization approach with Pareto-efficiency frontier – the "multi objective particle swarm" (MOPS) – to solve the MMJIT with setups was proposed through a test case of 20 different products production on 40 time buckets [47].

In [48], the authors compared a Bounded Dynamic Programming (BPD) procedure with GCM and with an Ant Colony (AC) approach, using as OF the minimization of the total inventory cost. They found that BDP is effective (1,03% as the average relative deviation from optimum) but not efficient, requiring roughly the triple of the time needed by the AC approach. Meanwhile, GCM was able to find the optimum (13% as the average relative deviation from optimum) on less than one third of the scenarios in which the AC was successful.

A broad literature survey on MMJIT with setups can be found in [49] while a comprehensive review of the different approaches to determine both kanban number and the optimal sequence to smooth production rates is present in [10].

6. Criticism on MMJIT problem solution

Assumed that time wastes are a clear example of MUDA in Lean Production [3], complex mathematical approaches which require several minutes to compute one optimal sequence for MMJIT [44] should be discarded, given that the time spent calculating new scheduling solutions does not add any value to products. On the other side, it is notable that MRP computation requires a lot of time, especially when it is run for a low-capacity process (in which CRP-MRP or capacitated MRPs are required). However, being MRP a look-ahead system which considers the demand forecasts, its planning is updated only at the end of a predefined "refresh period", not as frequently as it may be required in a non-leveled JIT context. MRP was conceived with the idea that, merging the Bill-Of-Materials information with inventory levels and requirements, the production manager could define a short-term work plan. In most cases, MRP is updated no more than every week; thus, an MRP run may also take one day to be computed and evaluated, without any consequences for the production plan. On the contrary, the situation in JIT environment evolves every time a product is required from downstream. While MRP assumes the Master Production Schedule forecasts as an input, in JIT nobody may know what is behind the curtain, minute by minute.

Indeed, while a perfect JIT system does not need any planning update – simply because in a steady environment (e.g. heijunka) the optimal sequence should always be almost the same, at least in the medium term – real-world short-term variations can deeply affect the optimality of a fixed schedule production. For instance, a one-day strike of transport operators in a certain geographical area can entirely stop the production of a subset of models, and the lack of a raw material for one hour can turn the best scheduling solution into the worst. On top of this, while MRP relies on its "frozen period", JIT is exposed to variability because is supposed to effectively react to small changes in the production sequence. However, some authors noticed that the JIT sequences [10, 48, 50] are not so resistant to demand changes, so a single variation in the initial plan can completely alter the best solution. This is particularly true when the required production capacity gets near to the available. Thus, developing algorithm for solving the MMJIT problem under the hypothesis of constant demand or constant product mix seems useless.

JIT was developed for manual or semi-automated assembly line systems, not for completely automated manufacturing systems. The flexibility of the JIT approach requires a flexible production environment (i.e. the process bottleneck should not be saturated) and this is not an easy condition to be reached in real industries. Consequently, despite the competence of its operations managers, even a big multinational manufacturer may encounter several problems in implementing JIT if a significant part of its supplier is made of small or medium-size enterprises (SMEs), which are naturally more exposed to variability issues. On top of this, differently from MRP – where the algorithm lies within a software and is transparent for users – in JIT the product sequencing is performed by the workforce and managed through the use of simple techniques, such as the heijunka box, the kanban board or other visual management tools, e.g. *andons*. Thus, any approach to organize JIT production should be easily comprehensible to the workers and should not require neither expert knowledge nor a supercomputer to be applied.

7. Using simulations to validate JIT heuristics

As it has been said, finding good solutions for the MMJIT problem with setups using an algorithmic approach may take too long and, on top of this, the solution can be vulnerable to product mix changes. Indeed, Kanban technique and GCM I methods are the most used approaches to manage JIT production thanks to their simplicity [44]. Some companies, where SMED techniques [51] failed to reduce setup times, use a modified version of the kanban FIFO board, in order to prevent setups proliferation. Thus, a simple batching process is introduced: when more than one kanban is posted on the board, the workstation operator shall not start the job on the first row but, on the contrary, chooses the job which allows the workstation to skip the setup phase. As an example, given the original job sequence A-B-A-C-A-B for a workstation, if the operator is allowed to look two positions ahead, he would process A-A-B-C-A-B, saving one setup time. In such situations, where setup times cannot be reduced under a certain value, rather than giving up the idea of adopting the Lean Production approach, heuristics can be developed and tested in order to obtain a leveled production even if coping with long setup times or demand variability.

The most common method to analyze and validate heuristics is through simulation. Several authors agree that simulation is one of the best ways to analyze the dynamic and stochastic behavior of manufacturing system, predicting its operational performance [52, 53, 54]. Simulating, a user can dynamically reproduce how a system works and how the subsystems interact between each other; on top of this, a simulation tool can be used as a decision support system tool since it natively embeds the *what-if* logic [55]. Indeed, simulation can be used to test the solutions provided by Genetic Algorithms, Simulated Annealing, Ant Colony, etc. since these algorithms handle stochasticity and do not assume determinism. Simulation can be used for:

- productivity analysis [56],
- production performances increase [1, 57, 58],
- confrontation of different production policies [59]
- solving scheduling problems [50, 60].

In spite of these potentialities, there seem to be few manufacturing simulation software really intended for industrial use, which go beyond a simple representation of the plant layout and modeling of the manufacturing flow. On top of some customized simulators – developed and built in a high-level programming language from some academic or research group in order to solve specific cases with drastic simplifying hypotheses – the major part of commercial software implements a graphical model-building approach, where experienced users can model almost any type of process using basic function blocks and evaluate the whole system behavior through some user-defined statistical functions [61]. The latters, being multi-purpose simulation software, require great efforts in translating real industrial processes logic into the modeling scheme, and it is thus difficult to "put down the simulation in the manufacturing process" [55]. Indeed, the lack of manufacturing archetypes to model building seems one of the most remarkable weakness for most simulator tools, since their presence could simplify the model development process for who speak the "language of business" [62]. Moreover, commercial simulators show several limitations if used to test custom heuristics, for example to level a JIT production or to solve a line-balancing problem: some authors report typical weaknesses in presenting the simulation output [63] or limited functionalities in terms of statistical analysis [64], on top of the lack of *user-friendliness*. For instance, most common commercial simulation software do not embed the most useful random distributions for manufacturing system analysis, such as the Weibull, Beta and Poisson distribution. When dealing with these cases, it is often easier to build custom software, despite it requires strong competences in operations research or statistics that have never represented the traditional background of industrial companies analysts [64].

In order to widespread simulation software usage among the manufacturing industry, some authors underline the need of a standard architecture to model production and logistics processes [65, 66, 67]. Literature suggested to focus on a new reference framework for manufacturing simulation systems, that implement both a structure and a logic closer to real production systems and that may support industrial processes optimization [68, 69].

Moreover, given hardware increased performances, computational workload of a simulation tool is not a problem anymore [70] and it seems possible to develop simulators able to run in less than one minute even complex instances. The complexity of a manufacturing model is linked both to size and system stochasticity. A careful analysis of time series can provide useful information to be included in the simulator, in order to model stochastic variables linked to machine failures or scrap production. This allows a more truthful assessment of key performance indicators (KPI) for a range of solutions under test.

8. Conclusions and a roadmap for research

The effective application of JIT cannot be independent from other key components of a lean manufacturing system or it can "end up with the opposite of the desired result" [71]. Specifically, leveled production (heijunka) is a critical factor. The leveling problem in JIT, a mixed-model scheduling problem, was formalized in 1983 and named MMJIT. Several numbers of solving approaches for MMJIT have been developed during the last decades. Most of them assume constant demand and product mix. Zero setup-times hypothesis has been removed only since 2000, and few approaches still cope with stochasticity. On top of this, these algorithms, although heuristic based, usually spend too much time in finding a good solution. Simplification hypotheses, operations research competences requirements and slow execution prevented these approaches to widespread in industry. Indeed, the heijunka box or the standard FIFO kanban approach with the simple Goal-Chasing-Method heuristic are still the most used tools to manage production in JIT environment. This is acknowledged also by the proponents of alternatives, and GCM is always used as a benchmark for every new MMJIT solution. However, these traditional approaches are not so effective in case of long setups and demand variations, given the fact that they have been conceived in pure JIT environments. In high stochastic scenarios, in order to prevent stock-outs, kanban number is raised along with the inventory levels. There are several cases of companies, operating in unstable contexts and where setup times cannot be reduced over a certain extent, that are interested in applying JIT techniques to reduce inventory carrying costs and manage the production flow in an effective and simple way. The development of kanban board / heijunka-box variations, in order to cope with the specific requirements of these companies, seems to offer better potentialities if compared to the development of difficult operations research algorithmic approaches. In order to solve industrial problems, researchers may concentrate in finding new policies that could really be helpful for production systems wishing to benefit from a JIT implementation but lacking in some lean production requirements, rather than studying new algorithm for MMJIT problem.

For instance, kanban board / heijunka-box variations can effectively focus on job preemption opportunities in order to reduce setups abundance, or on new rules to manage priorities in case of breakdowns or variable quality rates. The parameters fine-tuning can be performed through simulation. In this sense, given the limitations of most commercial software, the development of a simulation conceptual model – along with its requisites – of a model representation (objects and structures) and some communication rules between the subsystems (communication protocols) are the main issues that need to be addressed from academics and developers.

Author details

Francesco Giordano and Massimiliano M. Schiraldi

Department of Enterprise Engineering, "Tor Vergata" University of Rome, Italy

References

[1] J. P. Womack, D. T. Jones & D. Roos, The machine that changed the world: The Story of Lean Production, New York (NY): HarperPerennial, 1991.

[2] M. Rother & J. Shook, Learning to See: Value Stream Mapping to Add Value and Eliminate Muda, Brookline (MA): The Lean Enterprise Institute, 1999.

[3] T. Ohno, Toyota Production System: Beyond Large-Scale Production, Productivity Press, 1988.

[4] R. J. Schonberg, Japanese manufacturing techniques: Nine hidden lessons in simplicity, New York (NY): Free Press, 1982.

[5] J. Orlicky, Material Requirement Planning, New York (NY): McGraw-Hill, 1975.

[6] L. Baciarello, M. D'Avino, R. Onori & M. Schiraldi, « Lot-Sizing Heuristics Performance» working paper, 2013.

[7] A. Bregni, M. D'Avino e M. Schiraldi, «A revised and improved version of the MRP algorithm: Rev MRP,» *Advanced Materials Research (forthcoming)*, 2013.

[8] M. D'Avino, V. De Simone & M. Schiraldi, «Revised MRP for reducing inventory level and smoothing order releases: a case in manufacturing industry,» Production Planning & Control, 2013 (forthcoming).

[9] M. D'Avino, M. Correale & M. Schiraldi, «No news, good news: positive impacts of delayed information in MRP» working paper, 2013.

[10] C. Sendil Kumar & R. Panneerselvam, «Literature review of JIT-KANBAN system,» *International Journal of Advanced Manufacturing Technologies*, pp. 393-408, 2007.

[11] B. J. Berkley, «A review of the kanban production control research literature,» *Production and Operations Management*, vol. 1, n. 4, pp. 393-411, 1992.

[12] B. Sharadapriyadarshini & R. Chandrasekharan, «Heuristics for scheduling in a Kanban system with dual blocking mechanisms,» *European Journal of Operational Research*, vol. 103, n. 3, pp. 439-452, 1997.

[13] O. Kimura & H. Terada, «Design and analysis of pull system, a method of multistage production control,» *International Journal of Production Research*, n. 19, pp. 241-253, 1981.

[14] B. Hemamalini & C. Rajendran, «Determination of the number of containers, production kanbans and withdrawal kanbans; and scheduling in kanban flowshops,» *International Journal of Production Research*, vol. 38, n. 11, pp. 2549-2572, 2000.

[15] R. Panneerselvam, Production and Operations Management, New Delhi: Prentice Hall of India, 1999.

[16] H. Wang & H.-P. B. Wang, «Optimum number of kanbans between two adjacent workstations in a JIT system,» *International Journal of Production Economics*, vol. 22, n. 3, pp. 179-188, 1991.

[17] D. Y. Golhar & C. L. Stamm, «The just in time philosophy: a literature review,» *International Journal of Production Research*, vol. 29, n. 4, pp. 657-676, 1991.

[18] F. W. Harris, «How many parts to make at once,» *Factory, The Magazine of Management*, vol. 10, n. 2, pp. 135-136, 1913.

[19] R. B. Chase, N. J. Aquilano & R. F. Jacobs, Operations management for competitive advantage, McGraw-Hill/Irwin, 2006.

[20] R. Suri, «QRM and Polca: A winning combination for manufacturing enterprises in the 21st century,» Center for Quick Response Manufacturing, Madison, 2003.

[21] P. Ericksen, R. Suri, B. El-Jawhari & A. Armstrong, «Filling the Gap,» *APICS Magazine*, vol. 15, n. 2, pp. 27-31, 2005.

[22] J. Liker, The Toyota Way: 14 Management principles from the world's greatest manufacturer, McGraw-Hill, 2003.

[23] M. L. Spearman, D. L. Woodruff & W. J. Hopp, «CONWIP: a pull alternative to kanban,» *International Journal of Production Research*, vol. 28, n. 5, pp. 879-894, 1990.

[24] R. P. Marek, D. A. Elkins & D. R. Smith, «Understanding the fundamentals of kanban and conwip pull systems using simulation,» in *Proceedings of the 2001 Winter simulation conference*, Arlington (VA), 2001.

[25] R. Suri, Quick Response Manufacturing: A companywide approach to reducing lead times, Portland (OR): Productivity Press, 1998.

[26] M. M. Schiraldi, La gestione delle scorte, Napoli: Sistemi editoriali, 2007.

[27] S. Meissner, «Controlling just-in-sequence flow-production,» *Logistics Research*, vol. 2, p. 45.53, 2010.

[28] E. M. Goldratt, The Goal, Great Barrington, MA: North river press, 1984.

[29] M. Qui, L. Fredendall & Z. Zhu, «TOC or LP?,» *Manufacturing Engineer*, vol. 81, n. 4, pp. 190-195, 2002.

[30] D. Trietsch, «From management by constraints (MBC) to management by criticalities (MBC II),» *Human Systems Management*, vol. 24, pp. 105-115, 2005.

[31] A. Linhares, «Theory of constraints and the combinatorial complexity of the product-mix decision,» *International Journal of Production Economics*, vol. 121, n. 1, pp. 121-129, 2009.

[32] Y. Monden, Toyota Production System, Norcross: The Institute of Industrial Engineers, 1983.

[33] N. Boysen, M. Fliedner & A. Scholl, «Level Scheduling for batched JIT supply,» *Flexible Service Manufacturing Journal,* vol. 21, pp. 31-50, 2009.

[34] W. Kubiak, «Minimizing variation of production rates in just-in-time systems: A survey,» *European Journal of Operational Research,* vol. 66, pp. 259-271, 1993.

[35] Y. Monden, Toyota Production System, An Integrated Approach to Just-In-Time, Norcross (GA): Engineering & Management Press, 1998.

[36] J. Miltenburg, «A Theoretical Basis for Scheduling Mixed-Model Production Lines,» *Management Science,* vol. 35, pp. 192-207, 1989.

[37] W. Kubiak & S. Sethi, «A note on "level schedules for mixed-model assembly lines in just-in-time production systems",» *Management Science,* vol. 37, n. 1, pp. 121-122, 1991.

[38] G. Steiner & J. S. Yeomans, «Optimal level schedules in mixed-model, multilevel JIT, assembly systems with pegging,» *European Journal of Operational Research,* pp. 38-52, 1996.

[39] T. N. Dhamala & S. R. Khadka, «A review on sequencing approaches for mixed-model just-in-time production systems,» *Iranian Journal of Optimization,* vol. 1, pp. 266-290, 2009.

[40] M. S. Akturk & F. Erhun, «An overview of design and operational issues of kanban systems,» *International Journal of Production Research,* vol. 37, n. 17, pp. 3859-3881, 1999.

[41] P. R. McMullen & P. Tarasewich, «A beam search heuristic method for mixed-model scheduling with setups,» *International Journal of Production Economics,* vol. 96, n. 2, pp. 273-283, 2005.

[42] F. Mooeni, S. M. Sanchez & A. J. Vakharia, «A robust design methodology for Kanban system design,» *International Journal of Production Research,* vol. 35, pp. 2821-2838, 1997.

[43] G. N. Krieg & H. Kuhn, «A decomposition method for multi-product kanban systems with setup times and lost sales,» *IEE Transactions,* vol. 34, pp. 613-625, 2002.

[44] T. Tamura, S. Nishikawa, T. S. Dhakar e K. Ohno, «Computational Efficiencies of Goal Chasing, SA, TS and GA Algorithms to Optimize Production Sequence in a Free Flow Assembly Line,» in *Proceedings of the 9th Asia Pasific Industrial Engineering & Management Systems Conference,* Bali, 2008.

[45] K. Ohno, K. Nakashima & M. Kojima, «Optimal numbers of two kinds of kanbans in a JIT production system,» *International Journal of Production Research,* vol. 33, pp. 1387-1401, 1995.

[46] H. Aigbedo, «On bills of materials structure and optimum product-level smoothing of parts usage in JIT assembly systems,» *International Journal of Systems Science,* vol. 40, n. 8, pp. 787-798, 2009.

[47] A. Rahimi-Vahed, S. M. Mirghorbani e M. Rabbani, «A new particle swarm algorithm for a multi-objective mixed-assembly line sequencing problem,» *Soft computing*, vol. 11, pp. 997-1012, 2007.

[48] N. Boysen, M. Fliedner & A. Scholl, «Sequencing mixed-model assembly lines to minimize part inventory cost,» *Operational Research Spectrum*, pp. 611-633, 2008.

[49] A. Allahverdi, J. N. D. Gupta & T. Aldowaisan, «A review of scheduling research involving setup considerations,» *International Journal of Management Sciences*, vol. 27, pp. 219-239, 1999.

[50] P. Rogers & M. T. Flanagan, «Online simulation for real-time scheduling of manufacturing systems,» *Industrial Engineering*, pp. p. 37-40, 2000.

[51] S. Shingo, A revolution in manufacturing: The SMED system, Productivity Press, 1985.

[52] V. A. Hlupic, «Guidelines for selection of manufacturing simulation software,» *IIE Transactions*, vol. 31, n. 1, pp. 21-29, 1999.

[53] A. M. Law, Simulation modeling and analysis, Singapore: McGraw-Hill, 1991.

[54] J. Smith, «Survey of the use of simulation for manufacturing system design and operation,» *Journal of manufacturing systems*, vol. 22, n. 2, pp. 157-171, 2003.

[55] H. Berchet, «A model for manufacturing systems simulation with a control dimension.,» *Simulation Modelling Practice and Theory*, pp. p.55-57, 2003.

[56] A. Polajnar, B. Buchmeister & M. Leber, «Analysis of different transport solutions in the flexible manufacturing cell by using computer simulation,» *International Journal of Operations and Production Management*, pp. 51-58, 1995.

[57] P. Rogers & R. J. Gordon, «Simulation for the real time decision making in manufacturing systems,» in *Proceedings of the 25th conference on winter simulation*, Los Angeles (CA), 1993.

[58] P. Rogers, «Simulation of manufacturing operations: optimum-seeking simulation in the design and control of manufacturing systems: experience with optquest for arena,» in *Proceedings of the 34th conference on winter simulation: exploring new frontiers*, San Diego (CA), 2002.

[59] S. S. Chakravorty & J. B. Atwater, «Do JIT lines perform better than traditionally balanced lines,» International *Journal of Operations and Production Management*, pp. 77-88, 1995.

[60] R. Iannone & S. Riemma, «Proposta di integrazione tra simulazione e tecniche reticolari a supporto della progettazione operativa.,» Università di Salerno, Salerno, 2004.

[61] D. A. Van Beek, A. T. Hofkamp, M. A. Reniers, J. E. Rooda & R. R. H. Schiffelers, «Syntax and formal semantics of Chi 2.0,» Eindhoven University of Technology, Eindhoven, 2008.

[62] J. Banks, E. Aviles, J. R. McLaughlin & R. C. Yuan, «The simulator: new member of the simulation family,» *Interfaces*, pp. 21-34, 1991.

[63] A. M. Law & S. W. Haider, «Selecting simulation software for manufacturing applications: practical guidelines and software survey.,» *Industrial Engineering*, 1989.

[64] L. Davis & G. Williams, «Evaluating and Selecting Simulation Software Using the Analytic Hierarchy Process,» *Integrated Manufacturing Systems*, n. 5, pp. 23-32, 1994.

[65] D. A. Bodner & L. F. McGinnis, «A structured approach to simulation modeling of manufacturing systems,» in *Proceedings of the 2002 Industrial Engineering Research Conference*, Georgia, 2002.

[66] S. Narayanan, D. A. Bodner, U. Sreekanth, T. Govindaraj, L. F. McGinnis & C. M. Mitchell, «Research in object-oriented manufacturing simulations: an assessment of the state of the art,» *IIE Transactions*, vol. 30, n. 9, 1998.

[67] M. S. Mujtabi, «Simulation modeling of manufacturing enterprise with complex material. Information and control flows,» *International journal of computer integrated manufacturing*, vol. 7, n. 1, pp. 29-46, 1994.

[68] S. Robinson, «Conceptual modeling for simulation: issues and research requirements,» in *Proceedings of the 2006 Winter Simulation Conference*, Piscataway (NJ), 2006.

[69] C. Battista, G. Dello Stritto, F. Giordano, R. Iannone & M. M. Schiraldi, «Manufacturing Systems Modelling and Simulation Software Design: A Reference Model,» in *XXII DAAAM International World Symposium* , Vienna, 2011.

[70] A. M. Law & W. D. Kelton, Simulation Modelling and Analysis, New York: McGraw-Hill., 1991, pp. 60-80.

[71] S. Shingo, A study of the Toyota Production System, Productivity Press, 1989.

Chapter 7

The Important Role of Packaging in Operations Management

Alberto Regattieri and Giulia Santarelli

Additional information is available at the end of the chapter

1. Introduction

The chapter focuses on the analysis of the impact of packaging in Operations Management (OM) along the whole supply chain. The product packaging system (i.e. primary, secondary and tertiary packages and accessories) is highly relevant in the supply chain and its importance is growing because of the necessity to minimize costs, reduce the environmental impact and also due to the development of web operations (i.e. electronic commerce).

A typical supply chain is an end-to-end process with the main purpose of production, transportation, and distribution of products. It is relative to the products' movements normally from the supplier to the manufacturer, distributor, retailer and finally the end consumer. All products moved are contained in packages and for this reason the analysis of the physical logistics flows and the role of packaging is a very important issue for the definition and design of manufacturing processes, improvement of layout and increase in companies' efficiency.

In recent years, companies have started to consider packaging as a critical issue. It is necessary to analyse the packages' characteristics (e.g. shape, materials, transport, etc.) in order to improve the performance of companies and minimize their costs. Packaging concerns all activities of a company: from the purchasing of raw materials to the production and sale of finished products, and during transport and distribution.

In order to manage the activities directly linked with the manufacturing of products (and consequently with the packaging system), the OM discipline is defined. It is responsible for collecting various inputs and converting them into desired outputs through operations [1].

Recently, more and more companies have started to use web operations. Electronic commerce (e-commerce) is the most promising application of information technology witnessed in recent years. It is revolutionising supply chain management and has enormous potential for manufacturing, retail and service operations. The role of packaging changes with the increase in the use of e-commerce: from the traditional "shop window" it has become a means of information and containment of products.

The purpose of the chapter is to briefly describe a model of OM discipline usable to highlight the role of packaging along the supply chain, describing different implications of an efficient product packaging system for successful management of operations. Particular attention is paid to the role of product packaging in modern web operations.

The chapter is organised as follows: Section 2 presents a brief description of OM in order to engage the topic of packaging. The packaging logistics system is described in Section 3, before presenting experimental results of studies dealing with packaging perception by both companies and customers [2; 3]. Moreover, Section 3 introduces the packaging logistics system also including the analysis of the role of packaging in OM and a description of a complete mathematical model for the evaluation of total packaging cost is presented. Section 4 presents background about modern e-commerce and its relationship with OM. Packaging and e-commerce connected with OM is described in Section 5 and a case study on packaging e-commerce in operations is analysed in Section 6. Finally, the conclusion and further research are presented.

2. Operations management in brief

The brief introduction to OM wants to introduce the important role of packaging in all activities of a company. This section will describe a model of OM discipline that the authors have taken as a reference for dealing with the packaging topic.

According to Drejer et al. [4], the "Scientific Management" approach to industrial engineering developed by Frederick Taylor in the 1910s is widely regarded as the basis on which OM, as a discipline, is founded. This approach involved reducing a system to its simplest elements, analysing them, and calculating how to improve each element.

OM is the management function applied to manufacturing, service industries and no-profit organizations [5] and is responsible for all activities directly concerned with making a product, collecting various inputs and converting them into desired outputs through operations [1]. Thus, OM includes inputs, outputs, and operations. Examples of inputs might be raw materials, money, people, machines, and time. Outputs are goods, services, staff wages, and waste materials. Operations include activities such as manufacturing, assembly, packing, serving, and training [1]. The operations can be of two categories: those that add value and those with no added value. The first category includes the product processing steps (e.g. operations that transform raw materials into good products). The second category is actually a kind of waste. Waste consists of all unnecessary movements for completing an operation, which should therefore be eliminated. Examples of this are waiting time, piling products, reloading, and movements. Moreover, it is important to underline, right from the start, that the packaging system can represent a source of waste, but at the same time, a possible source of opportunities. Before waste-to-energy solutions, for example, it is possible to consider the use of recycled packages for shipping products. The same package may be used more than once; for example, if a product is sent back by the consumer, the product package could be used for the next shipment.

OM can be viewed as the tool behind the technical improvements that make production efficient [6]. It may include three performance aims: efficiency, effectiveness, and customer satisfaction. Whether the organization is in the private or the public sector, a manufacturing or non-manufacturing organization, a profit or a non-profit organization, the optimal utilization of resources is always a desired objective. According to Waters [1], OM can improve efficiency of an operation system to do things right and as a broader concept. Effectiveness involves optimality in the fulfilment of multiple objectives with possible prioritization within them; it refers to doing the seven right things well: the right operation, right quantity, right quality, right supplier, right time, right place and right price. The OM system has to be not only profitable and/or efficient, but must necessarily satisfy customers.

According to Kleinforfer et al. [7], the tools and elements of the management system need to be integrated with company strategy. The locus of the control and methodology of these tools and management systems is directly associated with operations. With the growing realization of the impact of these innovations on customers and profit, operations began their transformation from a "neglected stepsister needed to support marketing and finance to a cherished handmaiden of value creation" [8].

According to Hammer [9], a wave of change began in the 1980s called Business Process Reengineering[1] (BRP). BPR provided benefits to non-manufacturing processes by applying the efforts that Total Quality Management[2] (TQM) and Just In Time[3] (JIT) had applied to manufacturing. Gradually, this whole evolution came to be known as Process Management, a name that emphasized the crucial importance of processes in value creation and management. Process management is given further impetus by the core competency movement [13], which stressed the need for companies to develop technology-based and organizational competencies that their competitors could not easily imitate. The confluence of the core competency and process management movements led to many of the past decade's changes including the unbundling of value chains, outsourcing, and innovations in contracting and supply chains. People now recognize the importance of aligning strategy and operations, a notion championed by Skinner [14].

As companies developed their core competencies and included them in their business processes, the tools and concepts of TQM and JIT have been applied to the development of new products and supply chain management [7]. Generally, companies first incorporated JIT be-

---

1 BRP is the fundamental re-thinking and radical re-design of business processes to achieve improvements in critical contemporary measures of performance, such as cost, quality, service, and speed [10].

2 TQM is an integrative philosophy of management for continuously improving the quality of products and processes [11].

3 JIT is a manufacturing program with the primary goal of continuously reducing, and ultimately eliminating all forms of waste [12].

tween suppliers and production units. The 1980s' introduction of TQM and JIT in manufacturing gave rise to the recognition that the principles of excellence applied to manufacturing operations could also improve business processes and that organizations structured according to process management principles would improve. According to Kleindorfer et al. [7], the combination of these process management fundamentals, information and communication technologies, and globalization has provided the foundations and tools for managing today's outsourcing, contract manufacturing, and global supply chains. In the 1990s companies moved over to optimized logistics (including Efficient Consumer Response[4] (ECR)) between producers and distributors, then to Customer Relationship Management[5] (CRM) and finally to global fulfilment architecture and risk management. These supply chain-focused trends inspired similar trends at the corporate level as companies moved from lean operations to lean enterprises and now to lean consumption [17]. Figure 1 shows these trends and drivers, based on Kleindorfer et al. [7].

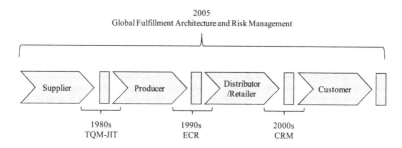

Figure 1. Trends and drivers of Operations Management (1980-2000) [7]

In order to manage the supply chain, organizations have to make different decisions about OM that can be classified as strategic, tactical, and operational. A graphical representation of the three decision levels of OM is shown in Figure 2 [18].

The three decisions' levels of OM interact and depend on each other: the strategic level is a prerequisite for the tactical level, and this in turn is a prerequisite for the operational level.

Strategic, tactical, and operational levels of OM are closely connected with the packaging system. Packaging is cross-functional to all company operations, since it is handled in several parts of the supply chain (e.g. marketing, production, logistics, purchasing, etc.). A product packaging system plays a fundamental role in the successful design and management of the operations in the supply chain. An integrated management of the packaging system from the strategic (e.g. decision of defining a new packaging solution), tactical (e.g. definition of the main packaging requirements) and operational (e.g. development of the physical

---

4 ECR is an attempt to increase the velocity of inventory in the packaged goods industry throughout the supply chain of wholesalers, distributors and ultimately end consumers [15].

5 CRM is a widely implemented model for managing company's interactions with customers. It involves using technology to organize, automate, and synchronize business processes [16].

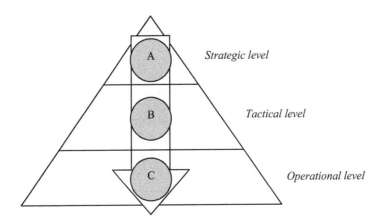

Figure 2. Graphical representation of OM's decision levels [18]

packaging system and respect of the requirements) point of view, allows companies to find the optimal management of the packaging system and to reduce packaging cost.

A general framework of product packaging and the packaging logistics system will be presented in Section 3.

2.1. Sustainable operations management

OM is increasingly connected with the environment and sustainable development (i.e. the development that meets the needs of the present without compromising the ability of future generations to meet their own needs), and it now concerns both the operational drivers of profitability and their relationship with people and the planet.

Following the definition of sustainability by the World Commission on Environment and Development (WCED), Sustainable Operations Management (SOM) is defined as *the set of skills and concepts that allow a company to structure and manage its business processes in order to obtain competitive returns on its capital assets without sacrificing the needs of stakeholders and with regard for the impact of its operations on people and environment.*

In order to perform sustainable operations, it is necessary to enlarge the perspective of OM, including people and the planet. According to Kleindorfer et al. [7], SOM integrates the profit and efficiency orientation of traditional OM with broader considerations of the company's internal and external stakeholders and its environmental impact. SOM helps companies to become agile, adaptive and aligned, balancing the people and the planet with profits [7].

Figure 1 has shown the evolution of OM since the 1980s. Figure 3 shows the impact of the SOM in the supply chain [7]. SOM has emerged over recent years and it influences the entire life cycle of the product (e.g. the management of product, recovery and reverse flows).

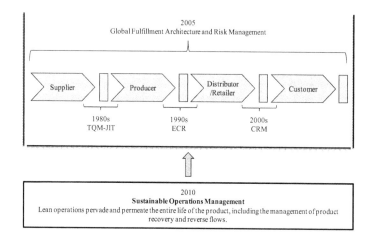

Figure 3. The impact of SOM in the supply chain (1980-2010) [7]

Considering the sustainability, environmental responsibility and recycling regulations, the packaging system plays an increasingly important role. Several environmental aspects are affected by packaging issues:

• Waste prevention: packages should be used only where needed. Usually, the energy content and material usage of the product being packaged are much greater than that of the package;

• Material minimization: the mass and volume of packages is one of the criteria to minimize during the package design process. The use of "reduced" packaging helps to reduce the environmental impacts;

• Re-use: the re-use of a package or its component for other purposes is encouraged. Returnable packages have long been used for closed loop logistics systems. Some manufacturers re-use the packages of the incoming parts for a product, either as packages for the outgoing product or as part of the product itself;

• Recycling: the emphasis focuses on recycling the largest primary components of a package: steel, aluminium, paper, plastic, etc.;

• Energy recovery: waste-to-energy and refuse-derived fuel in facilities are able to make use of the heat available from the packaging components;

• Disposal: incineration, and placement in a sanitary landfill are needed for some materials.

According to the studies conducted by Regattieri et al. [2; 3], users and companies have shown an interest in the environment and its link with the packaging system. Indeed, they believe that careful use of packaging can lead to an important reduction in environmental impact. Companies have begun to use recyclable materials (e.g. cardboard, paper, and plastic) and to re-use packages for other activities (for example online retailers are beginning to

re-use secondary packages of back products for future shipments). The next section describes the packaging system and its crucial role for the activities along the supply chain,
and then in OM.

3. A theoretical framework of the packaging system

During recent decades, the importance of the packaging system and its different functions
has been increasing. Traditionally, packaging is intended as a means of protecting and preserving goods, handling, transport, and storage of products [19]. Other packaging functions
like sales promotion, customer attention and brand communication have consistently grown
in importance [20]. It means that when a packaging developer makes a package, it needs to
be designed in order to meet the demand from a sales and a marketing perspective, and not
only from a manufacturing process and transportation network perspective [21].

The European Federation defines packaging *as all products made of any materials of any nature
to be used for the containment, protection, delivery and presentation of goods, from raw materials to
processed goods.*

Packaging is built up as a system usually consisting of a primary, secondary, and tertiary
level [22]. The primary package concerns the structural nature of the package; it is usually
the smallest unit of distribution or use and is the package in direct contact with the contents.
The secondary package relates to the issues of visual communication and it is used to group
primary packages together. Finally, the tertiary package is used for warehouse storage and
transport shipping [23].

A graphical representation of packaging system is shown in Figure 4:

The packaging system is cross-functional, since it interacts with different industrial departments, with their specific requests of how packages should be designed, and these are often
contradictory. Thus, packages have to satisfy several purposes, such as:

- Physical protection: the objects enclosed in the package may require protection from mechanical shock, vibration, electrostatic discharge, compression, temperature, etc.;

- Hygiene: a barrier from e.g. oxygen, water vapour, dust, etc. is often required. Keeping
the contents clean, fresh, sterile and safe for the intended shelf life is a primary function;

- Containment or agglomeration: small objects have to be grouped together in one package
for efficiency reasons;

- Information transmission: packages can communicate how to use, store, recycle, or dispose of the package or product;

- Marketing: packages can be used by marketers to encourage potential buyers to purchase
the product;

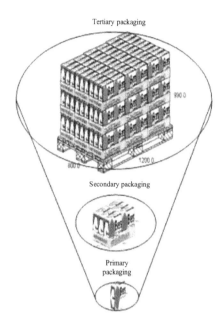

Figure 4. Graphical representation of the packaging system

• Security: packages can play an important role in reducing the risks associated with shipment. Organizations may install electronic devices like RFID tags on packages, to identify the products in real time, reducing the risk of thefts and increasing security.

3.1. Packaging system and operations management

In recent years, packaging design has developed into a complete and mature communication discipline [24]. Clients now realize that packages can be a central and critical element in the development of an effective brand identity. The packaging system fulfils a complex series of functions, of which communication is only one. Ease of processing and handling, as well as transport, storage, protection, convenience, and re-use are all affected by packaging.

The packaging system has significant implications in OM. In order to obtain successful management of operations, packaging assumes a fundamental role along the whole supply chain and has to be connected with logistics, marketing, production, and environment aspects. For example, logistics requires the packages to be as easy as possible to handle through all processes and for customers. Marketing demands a package that looks nice and is the right size. Packages do not only present the product on the shelf but they also arouse consumers' expectations and generate a desire to try out the product. Once the product is purchased, packages reassure the consumer of a product's quality and reinforce confidence [24]. Production requires only one size of packaging for all kinds of products in order to minimize time and labour cost. The environmental aspect demands the packaging system to be recyclable and to use the least material possible. Figure 5 shows the main interactions of the packaging system.

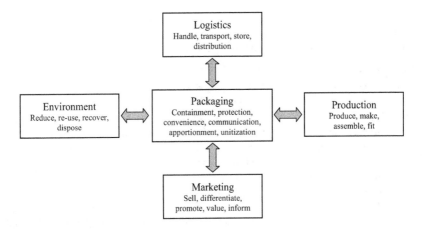

Figure 5. The main interactions of the packaging system

Scholars dealing with packaging disagree about its main function: some researchers emphasize that packaging is a highly versatile marketing tool [20], while others consider it mainly as an integral element of the logistics function [19; 25]. It is necessary to balance the technological and marketing aspects of packaging, indeed it has a significant impact on the efficiency of both logistics (e.g. manufacturing and distribution costs, time required for completing manufacturing and packing operations, which affect product lead time and due date performance to the customer) and the marketing function (e.g. products' selling, shelf presentation, etc.).

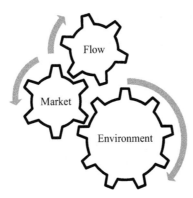

Figure 6. The main functions of the packaging system [26]

During the recent decades, the environmental aspect is considered by companies that deal with the packaging system. According to Johansson [26] the packaging system can be divided in three main functions, that interact each other: flow, market and environment (Figure 6).

The flow function consists of packaging features that contribute to more efficient handling in distribution. Packaging logistics, internal material flows, distribution, unpacking, disposal and return handling are included in this function.

Packaging logistics is a relatively new discipline that in recent years has been developed and has gained increasing attention in terms of the strategic role of logistics in delivering competitive advantage by the industrial and scientific community [22; 25]. Industry and science attribute different maturity levels to the subject depending on country and culture. According to Saghir [22], the concept of packaging logistics focuses on *the synergies achieved by integrating packaging and logistics systems with the potential of increased supply chain efficiency and effectiveness, through the improvement of both packaging and logistics related activities.* A more recent definition of packaging logistics is attributed to Chan et al. [27], who describe packaging logistics as the *interaction and relationship between logistics and packaging systems that improve add-on values on the whole supply chain, from raw material producers to end users, and the disposal of the empty package, by re-use, material recycling, incineration or landfill.* Both the definitions ([22; 27]) focus on the importance of the packaging logistics system, mainly in order to improve the efficiency of the whole supply chain.

In the market function, things like design, layout, communication, ergonomic aspects that create value for the product and the brand are important features for the packaging system [18]. The purpose of the market function is to satisfy customers and to increase product sales.

During recent decades the link between packaging and marketing is analysed in depth by several authors, and packaging has been studied as a marketing instrument that can influence some specific aspects, such as product positioning, consumer attention, categorization and evaluation, usage behaviour, intention to purchase or brand communication [28]. The aspect is significant since the package plays the role of an important interface between the brand owner and the consumer. The initial impression of product quality by the consumers is often judged by the impression of the package [29].

In the current operational environment, planning innovations must take into account not only marketing and logistics functions, but also a factor that is emerging as increasingly important: the environmental aspect. It aims to reduce the negative effects of the packaging system on the environment. Issues like the use of fewer inputs for the same outputs and the re-use of materials, facilitate the recycling of packaging [18]. Verruccio et al. [28] suggest that an increasing number of companies are choosing approaches that take care of the environmental aspects. It is further established that the design of the packaging system heavily influences the environmental aspect of activities in the supply chain [29; 30-32].

With regard to packaging logistics, the use of an appropriate packaging system (in terms of functions, materials, size and shape) can improve the management of operations [18]:

1. *Facilitate goods handling.* This function considers the following aspects:

2. a. Volume efficiency: this is a function of packaging design and product shape. In order to optimize the volume efficiency of a package, this function can be split into two parts, internal and external filling degree. The first regards how well the space within a package is utilized. When using standardized packages with fixed sizes, the internal filling degree might not always be optimal. The external filling degree concerns the fitting of the primary packages with secondary and of secondary with tertiary [7]. Packages that perfectly fill each other can eliminate unnecessary handling and the risk of damage, but it is important not to be too ambitious. Too much packaging may be too expensive, and there is a point where it is less costly to allow some damage than to pack for zero damage;

3. b. Consumption adaptation: the quantity of packages must be adapted to the consumption in order to keep costs low and not to tie unnecessary capital. Moreover it is desirable to have flexible packages and a high turnover of the packaging stock [7];

4. c. Weight efficiency: the package must have the lowest possible weight, because volume and weight limit the possible amount to transport. The weight is even more important when packages are handled manually [7];

5. d. Handleability: the packaging must be easy to handle for people and automatic systems working in the supply chain, and final customers [7]. According to Regattieri et al. [2; 3], the handleability is considered the most critical packaging quality attribute by Italian companies and users;

6. *Identify the product.* The need to trace the position of goods during transport to the final destination can be achieved in different ways, for example by installing RFID tags in packages. Thanks to this new technology, it is possible to identify the position of both packages and products in real time. This system leads to a reduction in thefts, increase in security, mapping of the path of products and control of the work in progress;

7. *Protect the product.* The protection of the product is one of the basic functions of packaging for both companies and users [2; 3]. An unprotected product could cause product waste, which is negative from both the environmental and the economic point of view. Packages must protect products during manufacturing and assembly (within the factory), storage and picking (within the warehouse) and transport (within the vehicle) from surrounding conditions, against loss, theft and manipulation of goods.

3.2. The role of packaging along the supply chain

Due to the different implications of the packaging system with all the activities of an organization, as underlined in the previous paragraphs, packaging has to be considered an important competitive factor for companies to obtain an efficient supply chain.

The packaging function assumes a crucial role in all activities along the supply chain (e.g. purchase, production, sales, transport, etc.). It is transversal to other industrial functions such as logistics, production, marketing and environmental aspects. The packaging function has to satisfy different needs and requirements, trying to have a trade-off between them. Considering the simplified supply chain of a manufacturing company (Figure 7), it is possible to analyse the role of the packaging function for all the parties of the supply chain.

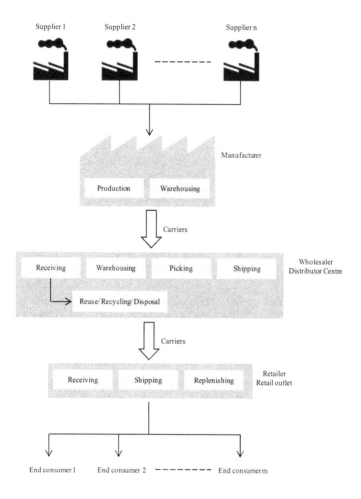

Figure 7. Typical supply chain of a manufacturing company

N suppliers provide raw materials to the manufacturer, which produces the finished products, sold to the distribution centre, then to the retailer and finally to m end consumers. In the middle, there are carriers that transport and distribute finished products along the supply chain. Each party has different interests and requirements regarding the function of packaging. Table 1 shows the different role of packaging for the parties to the supply chain.

Party	Role of packaging
n Suppliers	Suppliers are more interested in the logistics aspect of packaging than in marketing. They have to send products to the manufacturer and their purpose is the minimization of the logistics costs (transport, distribution, warehousing), so they prefer a package that is easy to handle and transport.
Manufacturer	The manufacturer produces finished products to sell to the distribution centre and, indirectly, to end consumers. It is important for the manufacturer to take into account all aspects: • product protection and safety, • logistics, • marketing and the • environment. Product protection and safety: the packages have to protect and contain the product, withstanding mechanical shocks and vibrations; Logistics: the manufacturer has to handle, store, pick and transport the product to the distribution centre. He has to make primary, secondary and tertiary packaging that is easy to transport, minimizes logistics costs and improves the efficiency of the company; Marketing: the manufacturer has to sell its products to the distribution centre that in turn sells to the retailer and in turn to end consumers. The manufacturer is indirectly in contact with end consumers and has to make primary packaging (the package that the users see on the shelf) that can incite the consumer to buy that product instead of another one. As Pilditch [33] said, the package is a "silent salesman", the first thing that the consumer sees when buying a product; Environment: people are more and more careful about protecting the environment. The manufacturer has to study a package that minimizes the materials used and can be re-usable or recyclable. The manufacturer has to balance the aspects described above in order to obtain an efficient supply chain.
Wholesaler	The wholesaler purchases products from the manufacturer and transports them to the distribution centre. He is mainly interested in the logistics aspect of packages since the most important functions are warehousing, picking and shipping the products. The wholesaler needs a package that is easy to handle and transport rather than one with an attractive shape and design.
Retailer	The retailer has to sell products to end consumers and for this reason, needs to consider what interests the end consumers. Marketing and environmental aspects are important: marketing because the package is a "shop window" for the product; environment since people are careful about minimizing pollution preferring to buy products contained in recyclable or re-usable packages.
m End consumers	End consumers are interested in marketing (indeed primary and secondary packages are effective tools for marketing in real shops [33]) and environmental aspects.

Table 1. The role of packaging for the parties along the supply chain

In conclusion, the packaging system plays a fundamental role along the entire supply chain where the parties often have opposite requirements and needs. Its design can be considered an element of OM discipline and must be integrated in the product design process taking into account logistics, production, marketing and environmental needs.

3.3. The perception of packaging by Italian companies and consumers [2; 3]

Regattieri et al. [2; 3] conducted two studies about the perception of packaging by Italian companies and users. The first deals with how Italian companies perceive and manage the packaging system, while the second discuss how Italian users perceive packaging quality attributes. The next two paragraphs briefly present the analysis conducted.

3.3.1. Packaging perception by Italian companies [2]

The study conducted by Regattieri et al. [2] is based on an explorative study of packaging development and packaging logistics, conducted in several Italian companies, from different industrial sectors. After the analysis of the Italian situation, the findings have been compared with the corresponding situation in Sweden. The comparison is mainly based on previous research conducted at the packaging logistics division of Lund University [34; 35].

In order to discuss the Italian industrial situation in terms of the packaging system, the authors implemented a questionnaire on packaging and its relationship with logistics, product and the environment. The quantitative content analysis of questionnaires allowed the authors to look in more depth at the Italian situation concerning packaging.

The first interesting data to underline is that more than half of companies (52.1%) think that packaging and its functions are critical and that their sales even depend on packaging (52.2%).

Another interesting analysis relates to packaging functions: protection and containment of the product are considered the most relevant function of packaging since it has effects on all activities throughout the supply chain, followed by product handling and communication (Figure 8). Like Italian companies, the packaging function most frequently mentioned by Swedish industries is the protection of products [34].

In order to obtain significant results on the product handling function, it is necessary to co-design product and packaging development. Companies are aware of the importance of integrating the development of the product with the development of the package: although a large percentage of Italian companies think the integration packaging and product is important and could reduce costs during the product life cycle, only 34.8% of them develop the packaging and the product at the same time. Italian companies, unlike Swedish ones, usually develop packaging after the designing the product.

In the same way as Swedish industries [34], Italian companies also consider logistics and transport an important packaging function. Indeed, 86.3% of companies report evaluating packaging costs from the transport point of view, mainly focusing on compatibility with vehicles and protection of goods (Figure 9). This data underlines the importance of the link between packaging and logistics systems: companies know that packaging (in terms of material, shape and size) influences storage, transport and distribution of goods. Although

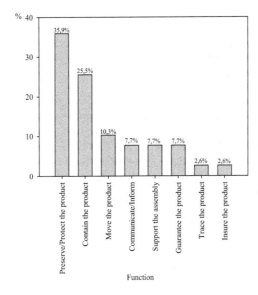

Figure 8. Classification of packaging functions

the most respondents compute packaging costs from the logistics point of view, only 39.1% of them report evaluating the total cost of packaging.

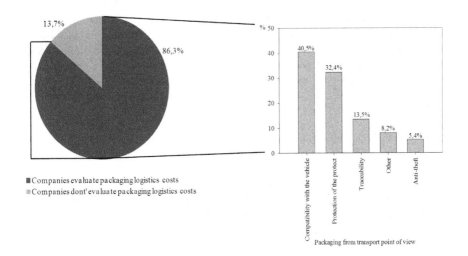

Figure 9. Classification of evaluating packaging logistics cost

The questionnaire also pointed out the importance of the relationship between packaging and the environment: 77.3% of Italian companies report using methods and applications in order to evaluate environmental aspects and 56.5% report recycling packaging materials. It is still a low percentage compared with Swedish data: in Sweden, consumer packages are largely recycled (e.g. 90% of glass, 73% of metal and 74% of paper and cardboard packages [36]).

The comparison between Italian and Swedish industries' perception of packaging has highlighted both Sweden's long-standing tradition in packaging development and in packaging logistics research and practice and the increasing attention of Italian industries on the importance of packaging functions (e.g. logistics and environmental aspects). Italian companies are following the Swedish ones in the development of a packaging logistics system and in the integration of packaging and product development, while maintaining their own characteristics. For more details, see Regattieri et al. [2].

3.3.2. Packaging perception by Italian customers [3]

The second analysis conducted by Regattieri et al. [3] is based on an explorative study conducted through a questionnaire distributed to Italian users. In order to understand how customer satisfaction may be increased, the authors analysed Italian consumers' perception of packaging quality attributes using the Theory of Attractive Quality, developed by Kano et al. in 1984 [37]. The findings are then compared with those of Swedish customers [38].

Kano et al. [37] defined a quality perspective in which quality attributes are divided into different categories, based on the relationship between the physical fulfilment of a quality attribute and the perceived satisfaction of that attribute. The five categories are attractive, one-dimensional, must-be, indifferent and reverse quality. All quality attributes can be satisfied or dissatisfied independently and they can change from one status to another according to the changes in customers' perspective. The packaging quality attributes are classified into three entities: *technical* (e.g. protection of the product, use of recyclable materials), *ergonomic* (everything relating to adaptations to human behaviour when using the product (e.g. ease of grip, ease of opening, user-friendly)) and *communicative* (the packaging's ability to communicate with customers (e.g. use of symbols, instructions for using packaging, brand communication)).

The questionnaire is made up of three parts:

• General information about the customers;

• Functional and dysfunctional question about packaging quality attributes. The classification into *attractive* (A), *one-dimensional* (O), *must-be* (M), *indifferent* (I), *reverse* (R) and *questionable* (Q) (Q responses include sceptical answers (Kano et al., 1984)) is made using an evaluation table (Figure 10), adapted by Löfgren and Witell [38] from Berger et al. [39].

• Level of importance of packaging characteristics: customers had to assign a value between 1 (not important) and 10 (very important) to the packaging quality attributes.

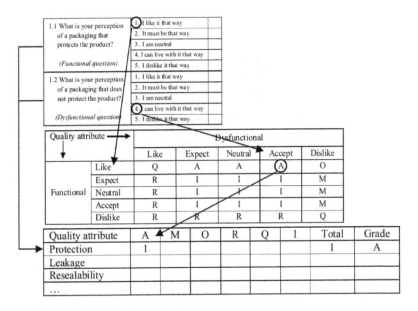

Figure 10. Evaluation table to classify packaging quality attributes (table adapted by [38] from [39])

The analysis of the questionnaires shows that Italian users are mainly interested in the ergonomic entity, made up of packaging characteristics that permit easy of handling of the product. Italians believe that the most important packaging function is protection of the product, according to the traditional role that has always been attributed to the packaging function.

For each packaging quality attribute, better and worse average values are calculated, indicating whether customer satisfaction can be increased by satisfying a certain requirement (better) or whether fulfilling this requirement may merely prevent customer dissatisfaction (worse) [39].

$$\text{Better average} = \frac{\sum_{i=1}^{n}(A+O)}{\sum_{i=1}^{n}(A+O+M+I)} \; \forall j \qquad \text{Worse average} = \frac{\sum_{i=1}^{n}(M+O)}{\sum_{i=1}^{n}(A+O+M+I)} \; \forall j$$

i=1,...,n is the number of responses for each packaging quality attribute

j=1,...,m represents packaging quality attributes

Figure 11 shows the Worse-Better Diagram for Italian users.

The Worse-Better Diagram focuses on technical, ergonomic and communicative entities. Contrary to the ergonomic and communicative entities, it is not possible to identify a definite cluster for the technical group, since the packaging quality attributes are scattered in the diagram, moving from one-dimensional (e.g. recyclable materials) to indifferent (e.g. additional functions) to must-be (e.g. protection of the product). Ergonomic and communicative

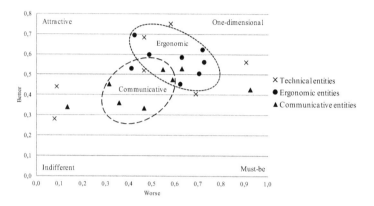

Figure 11. Worse-Better diagram for Italian perception on packaging quality attributes

entities assume definite clusters in the Worse-Better Diagram: the packaging quality attributes belonging to the ergonomic entity are mainly classified as one-dimensional. They are distinctive attributes that customers consider during the purchase of a product, comparing different brands. Italian customers locate the communicative quality attributes in the middle of the diagram. They delineate a specific cluster, but the dimension to which they belong is not clear.

Another important analysis is the level of importance attributed by Italian users to each packaging quality attribute. The highest values of importance are assigned to the protection of the product (9.59), open-dating (9.47), and hygiene (9.52). Italian customers seem to be interested neither in the aesthetics of packaging (attractive and nice looking print and the aesthetic appeal have low levels of importance: 4.52 and 5.00 respectively) nor in the additional functions (5.80).

From the comparison with the Swedish results [38], both Italians and Swedes have similar behaviour in terms of perception of packaging quality attributes. They consider the ergonomic quality characteristics the most significant packaging attributes, and the protection of the product the most important packaging function. Italians also perceive the use of recyclable material another important packaging attribute, in line with the growing importance of environmental considerations. Neither Italians nor Swedes place importance on aesthetics. For more details, see Regattieri et al. [3].

3.4. A mathematical model for packaging cost evaluation

As the previous paragraphs have underlined, the packaging system has numerous implications along the supply chain (e.g. marketing, production, logistics, purchasing, etc.). In order to define optimal management of the packaging system, it is necessary to evaluate the total packaging cost, made up of e.g. purchasing cost, manufacturing cost, transport and labour cost, management cost, etc. The study conducted by Regattieri et al. [2] underlines that most companies do not estimate the total packaging costs and, to confirm this, literature analysis shows the lack of a complete function for calculating the total cost of packaging in a company. For this reason, the authors have developed a complete mathematical model, considering all the cost parameters regarding the packaging system (primary, secondary and tertiary packages and accessories) along the whole supply chain of a manufacturing company.

The model represents added value for companies seeking to estimate the total costs of their packaging system and consequently its impact on total company costs. Moreover, it may be possible to find out the overlooked and oversized packaging factors. The former should be introduced in the calculation of the total packaging costs, while the latter could be reduced or eliminated.

Figure 12 shows the simplified supply chain of a manufacturing company.

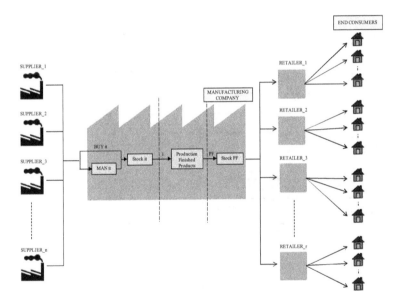

Figure 12. Simplified supply chain of a manufacturing company

The manufacturing company can rent or purchase packages (primary, secondary and tertiary and accessories) and raw materials (if the manufacturer produces packages internally) from the supplier n. When goods arrive, they are received in the manufacturer's receiving area, sorted and stored in the warehouse. If the company has to produce the packaging, the raw materials are picked and brought to the manufacturing area, where packages are made and subsequently stored in the warehouse. The raw materials not used during the manufacturing stage are brought back to the warehouse, creating a reverse flow of materials. When the finished products are produced, the packages are picked from the warehouse and brought to the manufacturing area. The packages not used during the manufacturing stage are brought back to the warehouse, creating a reverse flow of materials. The finished products are packed, put onto a pallet, and delivered to the retailer m. The model considers the possibility to re-use packages after the delivery of the finished products to the final customers and the possible disposal of packages if they are damaged. In addition, the model considers the possibility for the manufacturer to make a profit from sub-products derived from the disposal of packages and/or from the sale of tertiary packages to the final customers.

Table 2, 3 and 4 describe the indices, variables and cost parameters used in the model.

Index	Domain	Description
i	1,...,4	Level of package: i=1 (primary package) i=2 (secondary package) i=3 (tertiary package) i=4 (accessories)
t	1,...,m	Different packages for each level i
n	1,...,s	Suppliers
r	1,...,q	Retailers

Table 2. Indices of the model

Variable	Units	Description	Domain
x_{nit}	[pieces/year]	Quantity of raw materials bought by the company from the supplier n to produce package i of type t.	$i=1,...,4$; $t=1,...,m$; $n=1,...,s$
x'_{it}	[pieces/year]	Quantity of package i of type t produced by the manufacturer company from raw materials.	$i=1,...,4$; $t=1,...,m$
y_{nit}	[pieces/year]	Quantity of package i of type t bought by the company from supplier n.	$i=1,...,4$; $t=1,...,m$; $n=1,...,s$
w_{nit}	[pieces/year]	Quantity of package i of type t rented by the company from the supplier n.	$i=1,...,4$; $t=1,...,m$; $n=1,...,s$
r_{it}	[pieces/year]	Quantity of disposed package i of type t from which the	$i=1,...,4$; $t=1,...,m$

Variable	Units	Description	Domain
		company has a profit from sub-products.	
u_{rit}	[pieces/year]	Quantity of package i of type t sold by the company to the retailer r.	i=1,...,4; t=1,...,m; r=1,...,q
N_{ORD}	[orders/ year]	Number of orders for buying raw materials and/or packages i of type t.	
$N_{EXT\ TRAN\ nit}$	[trips/year]	Number of trips of raw materials and/or packages i of type t from the supplier n to the manufacturer.	i=1,...,4; t=1,...,m; n=1,...,s
$N_{INT\ TRAN\ it}$	[trips/year]	Number of trips of raw materials and/or packages i of type t from the manufacturer's receiving area to the warehouse.	i=1,...,4; t=1,...,m
$N_{INT\ TRAN}{}^{1}{}_{it}$	[trips/year]	Number of trips of raw materials i of type t from the warehouse to the manufacturing area to produce packages from x_{it}.	i=1,...,4; t=1,...,m
$N_{INT\ TRAN}{}^{2}{}_{it}$	[trips/year]	Number of trips of packages i of type t produced by the manufacturer and transported from the production area to the warehouse.	i=1,...,4; t=1,...,m
$N_{INT\ TRAN}{}^{3}{}_{it}$	[trips/year]	Number of trips of packages (produced/bought/rented) i of type t from the warehouse to the production area in order to support finished products.	i=1,...,4; t=1,...,m
$N_{REV\ INT\ TRAN}{}^{2}{}_{it}$	[trips/year]	Number of trips of packages i of type t not used during the production of finished products and transported from the manufacturing area to the warehouse.	i=1,...,4; t=1,...,m

Variable	Units	Description	Domain
$N_{REV\,INT\,TRAN}{}^1{}_{it}$	[trips/year]	Number of trips of the quantity of raw materials i of type t not used during the production of packages and transported from the manufacturing area to the warehouse.	i=1,...,4; t=1,...,m
$N_{REV\,EXT\,TRAN\,rit}$	[trips/year]	Number of trips of packages i of type t from the retailer r to the manufacturer.	i=1,...,4; t=1,...,m; r=1,...,q

Table 3. Variables of the model

Parameter	Nomenclatures	Units	Description
C_{ENG}	Cost of Engineering	[€/year]	Cost for studying each type of packaging and for making prototypes. It includes the labour costs of engineering the product.
C_{ORD}	Cost of Purchase Order	[€/order]	Cost for managing the internal purchase orders if the manufacturer produces the packaging internally; otherwise it represents the purchase orders for buying and/or renting packaging from suppliers. It includes the labour costs for making the order.
C_{PUR}	Cost of Purchasing	[€/piece]	Purchase cost of raw materials (to produce packaging) and/or packages.
C_{RENT}	Cost of Rent	[€/piece]	Cost to rent packages.
$C_{EXT\,TRAN}$	Cost of External Transport	[€/travel]	Cost for transporting raw materials and/or packages from the supplier to the manufacturer: it comprises labour costs, depreciation of vehicles (e.g. truck), cost of the distance travelled.
C_{REC}	Cost of Receiving	[€/year]	Cost for receiving raw materials and/or packages. It includes the labour costs and depreciation of vehicles (e.g. truck, forklift) used to unload products.
C_{COND}	Cost of Conditioning	[€/year]	Cost for sorting raw materials and/or packages before storing them in the warehouse. It includes the labour costs and depreciation of mechanical devices (if used), for example for unpacking and re-packing products.
$C_{INT\,TRAN}$	Cost of Internal Transport	[€/travel]	Cost for transporting raw materials and/or packages from the manufacturer's receiving area to the warehouse. It includes the labour costs, depreciation of vehicles (e.g. forklift), cost of the distance travelled.
C_{STOCK}	Cost of Stocking	[€/piece]	Cost for storing raw materials and/or packages in the warehouse. It includes the labour costs and the cost of the space for storing the packages.
C_{PICK}	Cost of Picking	[€/piece]	Cost for picking raw materials from the warehouse for producing the packages. It includes the labour costs and depreciation of vehicles (e.g. forklift) for picking the products.

Parameter	Nomenclatures	Units	Description
$C_{INT\,TRAN}$ [1]	Cost of Internal Transport[1]	[€/travel]	Cost for transporting raw materials from the warehouse to the manufacturing area to produce the packages. It includes the labour costs, depreciation of vehicles (e.g. forklift), cost of the distance travelled.
C_{MAN}	Cost of Packages Manufacturing	[€/piece]	Cost for producing packages internally; it includes the labour costs, depreciation of production plants and utilities (e.g. electricity, water, gas, etc.).
C_{REV} [1]	Cost of Internal Reverse Logistics[1]	[€/travel]	Cost of transport for bringing the raw materials not used during manufacturing back to the warehouse. It includes: $C_{REV\,INT\,TRAN}$ [1]: the cost of transport for coming back to the warehouse. It comprises labour costs, depreciation of vehicles used (e.g. forklift), cost of the distance travelled; $C_{REV\,INT\,COND}$ [1]: the cost of conditioning packages to make them re-usable. It comprises the labour costs and depreciation of mechanical devices (if used), for example for unpacking and re-packing products.
$C_{INT\,TRAN}$ [2]	Cost of Internal Transport[2]	[€/travel]	Cost for transporting the packages produced by the company from the production area to the warehouse. It includes the labour costs, depreciation of vehicles (e.g. forklift), cost of the distance travelled.
C_{STOCK} [1]	Cost of Stocking[1]	[€/piece]	Cost for stocking packages produced internally by the company. It includes the labour costs and cost of the space for storing the packages.
C_{PICK} [1]	Cost of Picking[1]	[€/piece]	Cost for picking packages (produced/bought/rented) from the warehouse. It includes the labour costs and depreciation of vehicles (e.g. forklift) for picking the packages.
$C_{INT\,TRAN}$ [3]	Cost of Internal Transport[3]	[€/travel]	Cost for transporting packages from the warehouse to the manufacturing area. It includes the labour costs, depreciation of vehicles (e.g. forklift), cost of the distance travelled.
C_{REV} [2]	Cost of Internal Reverse Logistics[2]	[€/travel]	Cost of transport for bringing packages not used during the manufacturing of finished products back to the warehouse. It includes: $C_{REV\,INT\,TRAN}$ [2]: the cost of transport for coming back to the warehouse. It comprises the labour costs, depreciation of vehicles used, cost of the distance travelled; $C_{REV\,INT\,COND}$ [2]: the cost of conditioning packages to make them re-usable. It comprises the labour costs and depreciation of mechanical devices (if used), for example for unpacking and re-packing products.
$C_{RE\text{-}USE}$	Cost of Re-Use	[€/year]	Cost of re-using packaging after the delivery of finished products to the customer. It includes: $C_{REV\,EXT\,TRAN}$: the cost of transport for coming back to the company. It comprises the labour costs, depreciation of vehicles used (e.g. truck), cost of the distance travelled; $C_{REV\,EXT\,COND}$: the cost of conditioning packages to make them re-usable. It comprises the labour costs and depreciation of mechanical devices (if used), for example for unpacking and re-packing products.
C_{DISP}	Cost of Disposal	[€/piece]	Cost of disposing of damaged packages during the manufacturing stage. It comprises the cost of disposal, the cost of transporting damaged packages from the company to the landfill (labour costs, depreciation of vehicles used (e.g. truck), cost of the distance travelled).

Parameter	Nomenclatures	Units	Description
R_{SUB}	Gain from Sub-Product	[€/piece]	The parameter identifies the possible gain obtained from the disposal of damaged products.
R_{UDC}	Gain from Direct Sale of Pallet	[€/piece]	This parameter identifies the possible gain obtained from the sale of tertiary packaging to the final customer.

Table 4. Cost parameters of the model

Equation (1) introduces the general formula of the model.

$$C_{TOT} = C_{ENG} + C_{ORD} + C_{PUR} + C_{RENT} + C_{EXT\ TRAN} + C_{REC} +$$
$$+ C_{COND} + C_{INT\ TRAN} + C_{STOCK} + C_{PICK} + C_{INT\ TRAN^1} + C_{MAN} + C_{REV^1} + \tag{1}$$
$$+ C_{INT\ TRAN^2} + C_{STOCK^1} + C_{PICK^1} + C_{INT\ TRAN^3} + C_{REV^2} + C_{RE\text{-}USE} + C_{DISP} - R_{SUB} - R_{UDC}$$

Equation (2) presents the mathematical model, explaining each cost parameter in detail.

$$C_{TOT} = \left(\sum_{i=1}^{4} \sum_{t=1}^{m} C_{ENG\ it} \right) + \left(N_{ORD} \cdot \sum_{i=1}^{4} \sum_{t=1}^{m} C_{ORD\ it} \right) + \left(\sum_{n=1}^{s} \sum_{i=1}^{4} \sum_{t=1}^{m} C_{PUR\ nit} \cdot (x_{nit} + y_{nit}) \right) +$$
$$+ \left(\sum_{n=1}^{s} \sum_{i=1}^{4} \sum_{t=1}^{m} C_{RENT\ nit} \cdot w_{nit} \right) + \left(\sum_{n=1}^{s} \sum_{i=1}^{4} \sum_{t=1}^{m} C_{EXT\ TRAN\ nit} \cdot N_{EXT\ TRAN\ nit} \right) +$$
$$+ \left(\sum_{i=1}^{4} \sum_{t=1}^{m} C_{REC\ it} \right) + \left(\sum_{i=1}^{4} \sum_{t=1}^{m} C_{COND\ it} \right) + \left(\sum_{i=1}^{4} \sum_{t=1}^{m} C_{INT\ TRAN\ it} \cdot N_{INT\ TRAN\ it} \right) +$$
$$+ \left(\sum_{i=1}^{4} \sum_{t=1}^{m} C_{STOCK\ it} \cdot (x_{it} + y_{it} + w_{it}) \right) + \left(\sum_{i=1}^{4} \sum_{t=1}^{m} C_{PICK\ it} \cdot x_{it} \right) +$$
$$+ \left(\sum_{i=1}^{4} \sum_{t=1}^{m} C_{INT\ TRAN^1 it} \cdot N_{INT\ TRAN^1 it} \right) + \left(\sum_{i=1}^{4} \sum_{t=1}^{m} C_{MAN\ it} \cdot x'_{it} \right) +$$
$$+ \left(\left(\sum_{i=1}^{4} \sum_{t=1}^{m} C_{REV\ INT\ TRAN^1 it} \cdot N_{REV\ INT\ TRAN^1 it} \right) + \left(\sum_{i=1}^{4} \sum_{t=1}^{m} C_{REV\ INT\ COND^1 it} \right) \right) + \tag{2}$$
$$+ \left(\sum_{i=1}^{4} \sum_{t=1}^{m} C_{INT\ TRAN^2 it} \cdot N_{INT\ TRAN^2 it} \right) + \left(\sum_{i=1}^{4} \sum_{t=1}^{m} C_{STOCK^1 it} \cdot x'_{it} \right) +$$
$$+ \left(\sum_{i=1}^{4} \sum_{t=1}^{m} C_{PICK^1 it} \cdot (x'_{it} + y_{it} + w_{it}) \right) + \left(\sum_{i=1}^{4} \sum_{t=1}^{m} C_{INT\ TRAN^3 it} \cdot N_{INT\ TRAN^3 it} \right) +$$
$$+ \left(\left(\sum_{i=1}^{4} \sum_{t=1}^{m} C_{REV\ INT\ TRAN^2 it} \cdot N_{REV\ INT\ TRAN^2 it} \right) + \left(\sum_{i=1}^{4} \sum_{t=1}^{m} C_{REV\ INT\ COND^2 it} \right) \right) +$$
$$+ \left(\left(\sum_{r=1}^{q} \sum_{i=1}^{4} \sum_{t=1}^{m} C_{REV\ EXT\ TRAN\ rit} \cdot N_{REV\ EXT\ TRAN\ riit} \right) + \left(\sum_{i=1}^{4} \sum_{t=1}^{m} C_{REV\ EXT\ COND\ it} \right) \right) +$$
$$+ \left(\sum_{i=1}^{4} \sum_{t=1}^{m} C_{DISP\ it} \right) - \left(\sum_{i=1}^{4} \sum_{t=1}^{m} R_{SUB\ it} \cdot r_{it} \right) - \left(\sum_{r=1}^{q} \sum_{i=1}^{4} \sum_{t=1}^{m} R_{UDC\ rit} \cdot u_{rit} \right)$$

The mathematical model allows companies to have a complete tool for analysing the total packaging costs in order to understand packaging cost reductions and consequently the minimization of the impact of total packaging cost on total company cost.

4. E-commerce

Among all operations, web operations are taking on an important role in the global trend of the purchasing process. During recent years, more and more people have begun to use the Internet and to buy a wide range of goods online. The World Wide Web (WWW) allows people to communicate simultaneously or asynchronously easily and effectively, shortening distance and time between individuals [40].

E-commerce is a new sales tool, in which consumers are able to participate in all the stages of a purchasing decision, while going through processes electronically rather than in a real shop. E-commerce is the process of trading goods, information, or services via computer networks including the Internet [41; 42]. There is an increasing consensus that e-commerce will represent a large share of retail markets in the future [43].

E-commerce channels in traditional companies have changed their operations and business strategy. That impact has been described by three main issues: integration, customization, and internationalization. First, e-commerce networks improve value chain integration by reducing transaction costs, facilitating JIT delivery, and improving information collection and processing [41; 42]. Secondly, e-commerce databases and direct links between producers and customers support high levels of product and service customization [44]. Finally, the Internet's international scope allows small companies to reach customers worldwide [45; 46].

As the Internet becomes more popular, e-commerce promises to become a mainstay of modern business [47]. There are dozens of e-commerce applications such as home banking, shopping in online stores and malls, buying stocks, finding a job, conducting an auction and collaborating electronically on research and development projects [42].

According to Gunasekaran et al. [48], e-commerce supports functional activities in organization: marketing, purchasing, design production, sales and distribution, human resource management, warehousing and supplier development. For example, the advent of e-commerce has changed marketing practice [48]. E-commerce systems should provide sure access to use, overcoming differences in time to business, location, and language between suppliers and customers and at the same time support the entire trading process in Business to Business (B2B) e-commerce [49]. Communication and data collection constraints are reduced with web-based production of goods and services. Using database management, data warehouse, and data mining technologies, the web can facilitate interaction with customers and suppliers, data collection, and data analysis processes [50].

Table 5 [48] summarises e-commerce applications and e-commerce tools and systems to suggest how e-commerce might support functional activities.

The open standard of the Internet ensures that large organizations can easily extend their trading communities, by increasing the efficiency of their business operations. According to Gunasekaran et al. [48], Internet-based e-commerce enables companies to:

- Shorten procurement cycles through the use of online catalogues, ordering, and payment;

Functional areas	E-commerce applications	E-commerce tools and systems
Marketing	Product promotion, new sales channels, direct savings, reduced cycle time, customer services.	B2B e-commerce, Internet ordering, website for the company.
Purchasing	Ordering, fund transfer, supplier selection.	EDI, Internet-purchasing.
Design	Customer feedback, research on customer requirements, product design, quality function deployment, data mining and warehousing.	WWW integrated CAD, Hyperlinks, 3D navigation, Internet for data and information exchange.
Production	Production planning and control, scheduling, inventory management, quality control.	B2B e-commerce, MRP, ERP, SAP.
Sales and distribution	Internet sales, selection of distribution channels, transportation, scheduling, third party logistics.	Electronic funds transfer, bar-coding system, ERP, WWW integrated inventory management, Internet delivery of products and services.
Human resource management	E-recruitment, benefit selection and management, training and education using WWW.	E-mails, interactive web sites, WWW based multimedia applications.
Warehousing	Inventory management, forecasting, scheduling of work force.	EDI, WWW integrated inventory management.
Supplier development	Partnership, supplier development.	WWW assisted supplier selection, e-mails, research on suppliers and products with WWW and intelligent agents.

Table 5. E-commerce applications areas, tools and systems [48]

- Reduce development cycles and accelerate time-to-market through collaborative engineering, product, and process design;
- Gain access to worldwide markets at a fraction of traditional costs;
- Significantly increase the speed of communication, especially international communication;
- Drastically reduce purchasing and production cycles;
- Reduce the cost of communication that in turn can reduce inventory and purchasing costs;
- Promote a closer relationship with customers and suppliers;
- Provide a quick and easy way of exchanging information about a company and its products, both internally and outside the organization.

5. Packaging and e-commerce in operations management

Every year Internet-based companies ship millions of packages throughout the world [24].

Online shopping influences packaging and its interactions with industrial function, mainly with marketing. The more people shop online, the more the role and the function of packaging change, since the shelf presentation of the product becomes less important [24]. Visser [24] stated that it is difficult to translate the existing packaging design used for the traditional way of buying in a real shop and marketing tactics into online retailing. E-commerce requires a new paradigm for the entire product packaging system. For example, in real shop the traditional primary package is a good agent for any products, not only because of the text descriptions, but also for its visual communication. It can effectively deliver product information and brand identity, and is a good cognitive agent for recognition. In an online shop, users cannot directly see the package nor touch the product, but other characteristics such as protection and re-usability for efficient take-back of products take on great importance [40]. The direct feeling with customers is less important since the contact is mediated by the computer.

The Internet does not determine the design of packages. However, if online shopping is becoming more common, packaging design must be reconsidered [24]. The changing role of packaging in the purchase of a product makes it desirable and possible to give more attention to the consumer's perception of a brand while the user is using it, and less attention to its shelf presentation. Retailers that sell online have to consider packages as a means of marketing and disseminating information instead of a mere covering for a product [24].

Block and Segev [51] suggest the following e-commerce impacts on marketing:

- Product promotion: e-commerce enhances the promotion of products and services through direct information and interactive contact with customers;

- New sales channels: e-commerce creates a new distribution channel for existing products, owing to its direct support of research on customers and the bidirectional nature of communication;

- Direct savings: the cost of delivering information to customers by Internet results in substantial savings. Greater savings are also made in the direct delivery of digitized products compared to the costs of traditional delivery;

- Reduced cycle time: the delivery time for digitized products and services can be reduced. Also, the administrative work related to physical delivery, especially across international borders, can be reduced significantly;

- Customer service: it can be greatly enhanced for customers to find detailed information online. In addition, intelligent agents can answer standard e-mail questions in few seconds.

The advent of e-commerce has also had several implications on logistics and the environment. From the logistics point of view, packaging has to increase its function of protection and covering of products, since products have to be transported to reach the customer. The theme of reverse logistics takes on great importance since customers can return wrong and/or unsuitable products. The advent of Internet distribution produces significant savings in shipping, and can facilitate delivery. Even those who use transportations can use Internet-based tools to increase customer service. Web-based order tracking has become commonplace. It allows customers to trace the shipment of their orders without having to contact the shipper directly [48]. Several electronic tools, like Electronic Data Interchange (i.e. the structured transmission of data between organizations by electronic means, EDI) can have a significant impact on the management of online packaging. EDI enables minimal stocks to be held with the consequent saving in storage, insurance, warehousing and labour costs (reduction in manual processing reduces the need for people) [48]. The packaging system must ensure secure shipping, reduce the possibility of theft, increase security and identify where the products are in real time.

From the environmental point of view, packaging in e-commerce has very similar requirements to traditional shopping, such as the use of recyclable materials, reduction of the amount of materials used, possibility to re-use packages in case of returned products from customers, disposal of damaged packages with the minimum production of pollution.

Table 6 shows the main interactions between packaging and other industrial issues in both real and online shopping.

Real shop	Online shop
Marketing:	Marketing:
Sell, differentiate, promote, value, inform, shelf presentation, visual communication	Brand identity, means of disseminating information, product promotion,
Logistics:	Logistics:
Handle, transport, store, distribution	Protection and covering the products, transport, reverse logistics, security
Environment:	Environment:
Reduction of materials used, re-use, recover, disposal	Reduction of materials, recyclable materials, re-use, disposal

Table 6. Packaging and industrial issues in real and online shops

6. A case study: Packaging e-commerce logistics in operations management

This section presents a case study on an Italian wholesaler; its main activities consist of purchasing goods from suppliers and selling and distributing them to retailers that in turn sell to end consumers through a "real shop".

The wholesaler is interested in starting a new business: the e-commerce activity. The wholesaler wants to sell directly to end consumers, bypassing the retailers and, at the same time, continue the B2B transactions.

Traditionally, the wholesaler receives goods from suppliers in the receiving area; the goods are unpacked, sorted and stored in the warehouse. When a retailer asks for products, they are picked from the shelves and packed according to the retailer's order. After that, the products packed in secondary packages are loaded onto the truck and dispatched to the retailer. Finally, he sells the products to end consumers in real shops. The packages are not labelled with identification technology (e.g. barcodes, RFID, etc.). Figure 13 shows in detail the activities of the wholesaler.

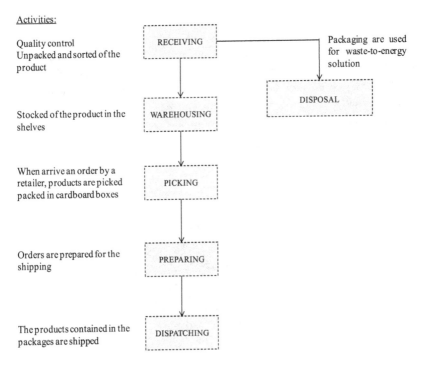

Figure 13. The wholesaler's activities

The project concerns the study of a new packaging system (in terms of material, shape, accessories used for protecting the product) to be used for online shopping. The new package has to take into account mainly the logistics aspects required by the e-commerce business.

The wholesaler has defined several requirements for the new packaging solution:

- Protection of the product: products contained in secondary packages have to be protected from mechanical shocks, vibrations, electrostatic discharge, compression, etc.;

- Handleability: the ergonomic aspect, that is everything relating to adaptations to the human physique and behaviour when using the product, has to be considered; the package has to be easy to open, easy to grip and user-friendly;

- Security: packages must ensure secure shipping. It is necessary to install identification technologies, like RFID tags or barcodes, in secondary packages in order to reduce thefts, increase security, and reduce costs and time spent on the traceability of products;

- Respect for the environment: the package has to be recyclable, in line with the requirements of end consumers and has to have minimum environmental impact;

- Re-use of packages from the supplier when the products back to the wholesaler.

The research activity starts from the study of several typical orders defined by the wholesaler in order to determine the best packaging configurations that optimize the combination of logistics, protection of the product and re-use of packages. The wholesaler decided to re-use the cardboard boxes in which the products are sent by suppliers. This solution minimizes the packaging system costs and reduces the environmental impact. According to these considerations, Figure 14 shows an example of the secondary package chosen.

Figure 14. The typical cardboard box used as secondary package

After that, the accessories are chosen in order to protect products from mechanical shocks, vibrations and compression during transport. Pluriball, polystyrene and interior cushioning are chosen as flexible protective accessories (an example of interior cushioning is shown in Figure 15).

The authors have analysed the possibility to install RFID tags on secondary packages in order to find out the position of the products in real time and to increase security during transport, minimizing the possibility of thefts and loss of products.

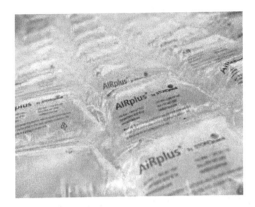

Figure 15. Accessories used for protecting products (courtesy of Soropack Group)

The new packaging solution presents several advantages in terms of:

- Protection of the product: the products inside the packages are protected thanks to the accessories used that increase the protection of products, damping the shocks during transport;

- Handleability: the package is user-friendly, easy to handle and to open;

- Security: the installation of RFID tags in the secondary packages allows the wholesaler to increase security during transport, reduce the number of thefts, and find out the position of the package at all times. This aspect may also be important for the end consumer since he can verify the position of the product he has ordered;

- Respect for the environment: the packages and accessories used for the e-commerce business can be recycled (the cardboard box is paper and the interior cushioning plastic) and secondary packages are re-used: the wholesaler use the cardboard with which the products arrive from the suppliers for dispatching products to end consumers.

In order to define a new packaging solution for the e-commerce business and, according to OM discipline, the strategic, tactical and operational levels have to be analysed. The definition of a new packaging solution for the e-commerce business, allowing transactions costs to be minimized and leading to an increase in business, is a strategic decision. The tactical management defines the main packaging requirements and the operational level has to implement the solution. The activities of the operational level are to test the products and packages in order to verify the resistance to shocks, build the website from sell by the WWW, study the shape, materials and accessories for packages, define a package that is as easy as possible to handle and transport and analyse the installation of RFID tags in secondary packages. Figure 16 shows in detail the decisions and operations at all levels in the pyramid of the OM decision levels.

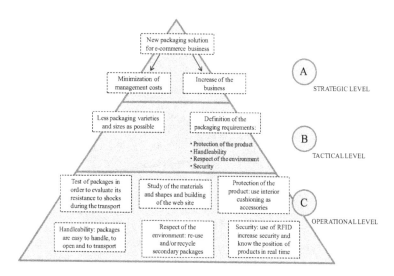

Figure 16. The pyramid of OM's decision levels for the case study

The new solution is implemented by the wholesaler and implies several benefits: an increase in sales with minimum effort, a reduction in transaction costs and an increase in customer satisfaction thanks to the environmentally friendly packaging. Moreover, the products are now traced every time and in real time, thanks to the installation of RFID tags in secondary packages, reducing thefts, loss and increasing security.

7. Conclusion

Operations Management is defined as the management function responsible for all activities directly concerned with making a product, collecting various inputs and converting them into desired outputs through operations [5]; OM discipline can be applied to manufacturing, service industries and non-profit organizations.

Over the years, new tools and elements such as TQM, JIT, and ECR have become part of the OM discipline that recognizes the need to integrate these tools and elements of the management system with the company's strategy. In order to manage all operations, organizations have to define a strategy, whose decisions are based on three levels: strategic, tactical and operational. Each level is integrated with the others and has to be interrelated in order to follow a common purpose. Strategic, tactical and operational decision levels are strictly connected with packaging features.

Packaging is a multidimensional function that takes on a fundamental role in organizations to achieve successful management of operations. Johansson [26] stated that the packaging system (made up of primary, secondary and tertiary packaging and accessories used to protect the products inside) could be divided into three main functions that interact with each other. They are flow, market and environment. The flow function consists of packaging features that contribute to more efficient handling during transport. The market function considers the aesthetics aspect in order to create value for the product and finally, the environment function has the purpose of reducing the negative effects of packaging on the environment. Packaging has an important role along the whole supply chain: all the parties (e.g. suppliers, manufacturers, retailers, end consumers) are interested in the packaging features (e.g. protection of the product, aesthetics aspects, reduction of the environmental impact, etc.).

In order to find the optimal packaging system management, the authors have developed a complete mathematical model that represents added value for companies seeking to estimate the total costs of their packaging system and consequently its impact on total company costs. The model considers all the cost parameters regarding the packaging system, e.g. engineering cost, warehousing cost, labour cost, transport cost, etc.

The packaging system takes on a fundamental role in online shopping. In recent years, web operations have evolved and organizations who want to start online business have to reconsider the role of packaging: from merely "shop window" in real shops, packaging has to transform into a means of information and transport. The changing role of packaging in the purchase of a product makes it desirable and possible to give more attention to the consumer's perception of a brand while he is using it, and less attention to its shelf presentation [24].

The correlation between packaging and e-commerce is a relatively new aspect. The case study described in Section 5 has shown the will of organizations to enter into the new e-commerce business, but also the changes that they have to make to the packaging system, since the packaging requirements of online shopping are different from those of a real shop. Organizations gain important benefits from e-commerce, such as the increase in labour cost savings.

Several modifications have to be considered for future thinking concerning online packaging. Communicative and information functions must be built in to help consumers to identify the products easily and to assist them in making precise decisions and reinforcing brand identity for consumers online. In addition, the ability to attract consumers' attention and incite their curiosity about the products are important points to analyse in the future in order to increase the potential development of packages for online shopping.

Author details

Alberto Regattieri[1] and Giulia Santarelli[2]

1 DIN – Department of Industrial Engineering, University of Bologna, Bologna, Italy

2 DTG – Department of Management and Engineering, University of Padova, Padova, Italy

References

[1] Waters D. Operations management – Producing goods and services. Addison-Wesley (eds.). Great Britain; 1996.

[2] Regattieri A., Olsson A., Santarelli G., Manzini R. An empirical survey on packaging perception for Italian companies and a comparison with Swedish situation. Proceedings of the 24th NOFOMA Conference, June 2012, Turku, Finland.

[3] Regattieri A., Santarelli G. and Olsson A. The Customers' Perception of Primary Packaging: a Comparison between Italian and Swedish Situations. Proceedings of the 18th IAPRI World Packaging Conference, June 2012, San Luis Obispo, California.

[4] Drejer A., Blackmon K., Voss C. Worlds apart? – A look at the operations management area in the US, UK and Scandinavia. Scandinavian Journal of Management 2000; 16 45-66.

[5] Waller D.L. Operations management: a supply chain approach. 2nd edition. Thompson (ed.). London; 2003.

[6] Schmenner R.W., Swink M.L. On theory in operations management. Journal of Operations Management 1998; 17 97-113.

[7] Kleindorfer P.R., Van Wassenhove L.N. Strategies for building successful global businesses. In: Gatignon and Kimberley (eds.) Managing risk in global supply chains. 2004. p288-305.

[8] Hayes R.H., Wheelwright S.C. Restoring out competitive edge: competing through manufacturing. In:Wiley, New York; 1984.

[9] Hammer M. Re-engineering work: don't automate, obliterate. Harvard Business Review 1990; 68(4) 104-112.

[10] Hammer M., Champy J. Reengineering the corporation: a manifesto for business revolution. National Bestseller, 1993.

[11] Ahire S.L. Total Quality Management interfaces: an integrative framework. Management Science 1997; 27(6) 91-105.

[12] Sugimori Y., Kusunoki F., Cho F., Uchikawa S. Toyota production system and kanban system: materialization of just-in-time and respect for human systems. International Journal of Production Research 1977; 15(6) 553-564.

[13] Hamel G., Prahalad C.K. Competing for the future: break-through strategies for sizing control of your industry and creating the markets of tomorrow. Harvard Business School Press (ed.). Boston, Massachussetts; 1994.

[14] Skinner W.S. Manufacturing strategy on the "S" curve. Production and Operations Management 1996; 5(1) 3-14.

[15] Coyle J.J., Bardi E.J., Langley C.J. Jr. The management of business logistics. West publishing company, St Paul, MN, 1996.

[16] Shaw R. Computer aided marketing & selling. In:Butterworth Heinemann; 1991.

[17] Womack J.P., Jones D.T. Lean consumption. Harvard Business Review 2005; 83(3) 58-68.

[18] Hansson E., Olsson M. Ellos: a case study in operations management and packaging logistics. School of economics and commercial low, Göteborg University, Sweden; 2000.

[19] Hellström D., Saghir M. Packaging and logistics interactions in retail supply chains. Packaging Technology and Science 2006; 20(3) 197-216.

[20] Underwood R.L. The communicative power of product packaging: creating brand identity via lived and mediated experience. Journal of Marketing Theory and Practice 2003; 11(1) 61-65.

[21] Silversson J., Jonson G. Handling time and the influence of packaging design. Licentiate Thesis Lund University, Sweden; 1998.

[22] Saghir M. Packaging logistics evaluation in the Swedish retail supply chain. PhD Thesis, Lund University, Sweden; 2002.

[23] Long D.Y. Commercial packaging design. In: Yellow Lemon. 1982, Taipei, Taiwan.

[24] Visser E. Packaging on the web: an underused resource. Design Management Journal 2002; 62-67.

[25] Twede D. The process of packaging logistical innovation. Journal of Business Logistics 1992; 13(1) 69-94.

[26] Johansson K., Lorenszon-Karlsson A., Olsmats C., Tiliander L. Packaging logistics. Packforsk, Kista; 1997.

[27] Chan F.T.S., Chan H.K., Choy K.L. A systematic approach to manufacturing packaging logistics. The International Journal of Advanced Manufacturing Technology 2006; 29(9;10) 1088-1101.

[28] Verruccio M., Cozzolino A., Michelini L. An exploratory study of marketing, logistics, and ethics in packaging innovation. European Journal of Innovation Management 2010; 13(3) 333-354.

[29] Olsson A., Larsson A.C. Value creation in PSS design through product and packaging innovation processes. In Sakao and Lindahl (eds.) Introduction to product/service-system design; 2009. p93-108.

[30] Nilsson F., Olsson, A., Wikström F. Toward sustainable goods flows – a framework from a packaging perspective. Proceedings of 23rd NOFOMA Conference, June 2001, Norway.

[31] Sonneveld K., James K., Fitzpatrick L., Lewis H. Sustainable packaging, how we define and measure it? 22nd IAPRI symposium of packaging; 2005.

[32] Svanes E. Vold M., Møller H., Kvalvåg Pettersen M., Larsen H., Hanssen O.J. Sustainable packaging design: a holistic methodology for packaging design. Packaging Technology and Science 2010; 23(2) 161–175.

[33] Pilditch J. The silent salesman. 2nd ed. In: Doble & Brendon (eds.). 1973, Plymouth.

[34] Bramklev C. A survey on the integration of product and package development. International Journal Manufacturing Technology and Management 2010; 19(3;4) 258-278.

[35] Bjärnemo R., Jönson G., Johnsson M. Packaging logistics in product development. In Singh J., Lew S.C. & Gay R. (eds.). Proceedings of the 5th International Conference: computer integrated manufacturing technologies for new millennium manufacturing, 2000, Singapore.

[36] Helander F. Svensk Förpacknignsindustri. Var är vi idag och vad påverkar utvecklingen framåt? Packbridge publication, Malmö, Sweden. 2010.

[37] Kano N., Seraku N., Takahashi F., Tsjui F. Attractive quality and must-be-quality. Hinshitsu 1984; 2 147-156.

[38] Löfgren M., Witell L. Kano's theory of attractive quality and packaging. The Quality Management Journal 2005; 12(3) 7-20.

[39] Berger C., Blauth R., Boger D., Bolster C., Burchill G., DuMouchel W., Poulist F., Richter R., Rubinoff A., Shen D., Timko M., Walden D. Kano's methods for understanding customer-defined quality. The Center of Quality Management Journal 1993; 2(4).

[40] Huang K.L., Rust C., Press M. Packaging design for e-commerce: identifying new challenges and opportunities for online packaging. College of Digital Design. Visual Communication Design Graduate School of Digital Content and Animation, 2009.

[41] Fraser J., Fraser N., McDonald F. The strategic challenge of electronic commerce. Supply Chain Management: An International Journal 2000; 5(1) 7-14.

[42] Turban E., Lee J., King D. and Chung H.M. Electronic commerce: a managerial perspective. Prentice-Hall International (UK) Limited, London, 2000.

[43] Giovani J.C. Towards a framework for operations management in e-commerce. International Journal of Operations & Production Management 2003; 23(2) 200-212.

[44] Skjoett-Larsen T. European logistics beyond 2000. International Journal of Physical Distribution & Logistics Management 2000; 30(5) 377-387.

[45] Soliman F., Youssef M. The impact of some recent developments in e-business in the management of next generation manufacturing. International Journal of Operations & Production Management 2001; 21(5;6) 538-564.

[46] Zugelder M.T. Flaherty T.B. and Johnson J.P. Legal issues associated with international internet marketing. International Marketing Review 2000; 17(3) 253-271.

[47] Altmiller J.C., Nudge B.S. The future of electronic commerce law: proposed changes to the uniform commercial code. IEEE Communication Magazine 1998; 36(2) 20-22.

[48] Gunasekaran A., Marri H.B., McGaughey R.E., Nebhwani M.D. E-commerce and its impact on operations management. International Journal of Production Economics 2002; 75 185-197.

[49] Boll S., Gruner A., Haaf A., Klas W. EMP – A database-driven electronic market place for business-to-business commerce on the internet. Distributed and Parallel Database 1999; 7(2) 149-177.

[50] Wang F., Head M., Archer N. A relationship-building model for the web retail marketplace. Internet Research 2000; 10(5) 374-384.

[51] Block M., Segev A. Leveraging electronic commerce for competitive advantage: a business value framework. Proceedings of the Ninth International Conference on EDI-ISO. 1996, Bled, Slovenia.

Chapter 8

Enterprise Risk Management to Drive Operations Performances

Giulio Di Gravio, Francesco Costantino and
Massimo Tronci

Additional information is available at the end of the chapter

1. Introduction

Global competition characterizes the market of the new millennium where uncertainty and volatility are the main elements affecting the decision making process of managers that need to determine scenarios, define strategies, plan interventions and investments, develop projects and execute operations (figure 1).

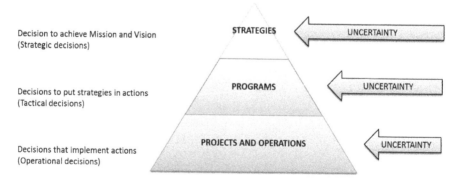

Figure 1. Decision hierarchy

Risks have been always part of entrepreneurships but a growing attention to the issues related to Risk Management is nowadays spreading. Along with the financial scandals in the affairs of some major corporations, the high degree of dynamism and the evolutions of markets need organizations to rapidly adapt their business models to changes, whether economic, political, regulatory, technological or social [1].

In particular, managerial trends of business disintegration, decentralization and outsourcing, pushed organizations towards practices of information sharing, coordination and partnership. The difficulties that generally arise during the implementation of these practices underline the impact that critical risk factors can have on corporate governance. Operations, at any level, are highly affected in their performance by uncertainty, reducing their efficiency and effectiveness while losing control on the evolution of the value chain.

Studies on risk management have to be extended, involving not only internal processes of companies but considering also the relationship and the level of integration of supply chain partners. This can be viewed as a strategic issue of operations management to enable interventions of research, development and innovation.

In a vulnerable economy, where the attention to quality and efficiency through cost reduction is a source of frequent perturbations, an eventual error in understanding the sensibility of the operations to continuous changes can seriously and irreparably compromise the capability of fitting customers' requirements.

Managers need to have personal skills and operational tools to ensure that risk management strategies can be suitably implemented and integrated in the production and logistics business environment. In order to face internal and external uncertainty, take advantage of it and exploit opportunities, it is necessary to identify, analyze and evaluate operational risks through standard methodologies that help to:

- classify the different types of risks;
- identify risks in scope;
- assess risks;
- identify possible interventions and relative priorities;
- select, plan and implement interventions, managing actions and collecting feedbacks.

While studies and standards on risk management for health and safety, environment or security of information defined a well-known and universally recognized state of the art, corporate and operational risk management already needs a systematic approach and a common view. The main contributions in these fields are the reference models issued by international bodies [2-5].

Starting from the most advanced international experiences, in this chapter some principles are defined and developed in a framework that, depending on the maturity level of organizations, may help to adequately support their achievements and drive operations performance.

2. Corporate governance and risk management

Over the years, the attention to the basic tenets of corporate governance has radically increased.

In response to the requirements of supporting business leaders in managing organizations and in protecting the various stakeholders towards the evolution of the political, economic and social environment, guide lines and reference models in the field of corporate governance have been issued.

Within this body of rules, risk management plays a main role. It relates directly to the recognition of the strategic connotations of corporate governance as the means to achieve business targets, according to the rights and expectations of stakeholders.

Since the mid-nineties onwards, the themes of risk management and corporate governance are strictly intertwined and almost coincident: the systematic management of risks has become a synonym of a "healthy" management of the business. At the same time, the techniques of risk analysis, historically associated with assessing financial risks, have been revised or replaced by methods that pervade the organization in depth. Along with the use of specific and complex control models (i.e. the experience of the Code of Conduct of the Italian Stock Exchange), responsibility for risk management is placed at the level of senior management. In some countries, such as Germany, Australia and New Zealand, these indications reached the level of compulsory requirements as national legislation asks all companies to have an operational risk management system.

From the above, the close link between corporate governance and risk management is absolutely clear. It has to be considered not only as an operational practice but rather as an essential component of decision making, based on the continuous development of definition systems and, therefore, of the top management responsibility.

The management of the company risk profile requires the knowledge of:

- the risk system affecting the enterprise;
- the nature and intensity of the different types of risks;
- the probability of occurrence of each risk and its expected impact;
- the mitigation strategies of the different types of risks.

To ensure that the approved, deliberated and planned risk management strategies are executed in an effective and efficient way, the company's top management shall periodically review and, if necessary, implement corrective and/or preventive action with regard to:

- reliability of existing systems for the identification and assessment of risks;
- effectiveness of internal control systems to monitor risks and their possible evolution.

Corporate governance is thus to be seen as the strategic platform on which the tactical and operational system of risk & control acts, i.e. the set of processes, tools and resources at all levels of the organization to ensure the achievement of corporate objectives. On these arguments, it is appropriate to consider that the application of a system based on the principles of risk & control governance allows the creation of a virtuous circle of performances that has a positive impact on the environment inside and outside the company, beyond regulatory requirements.

Management has the responsibility to plan, organize and direct initiatives to ensure the achievement of company goals, in terms of:

- definition of business and government targets;
- formulation of strategies to reach business and government targets;
- effective and efficient use of the resources of the organization;
- relevance and reliability of financial and operational reporting;
- protection of company assets;
- compliance with laws, regulations, contracts and corporate ethical standards;
- protection of ethical and social values.

The management acts through a regular review of its objectives, changes in processes according to changes in the internal and external environment, promoting and maintaining a business-oriented culture and a climate.

3. Risk classification

Uncertain events can have both a positive and a negative effect: on the one hand, in fact, they are a threat to the achievement of business objectives, on the other hand can become a significant source of opportunities for companies able to understand, anticipate and manage them. According to [6], risks are *"events with negative impacts that can harm the creation of business value or erode the existing one"* while *opportunities* are *"events with positive impact that may offset negative impacts"*. The opportunities are chances that an event will occur and positively affect the achievement of objectives, contributing thus to the creation of value or preserving the existing one. Management needs to assess the opportunities, reconsidering its strategies and processes of setting goals and developing new plans to catch benefits derived from them.

An inherent risk can so be defined as "the possibility that an event occurs and have a negative impact on the achievement of objectives" while the control can be defined as "any means used by management to increase the likelihood that the business objectives set are achieved", mitigating the risks in an appropriate manner. In this context, a hazard is a "potential source of risk" while a residual risk is the "risk that still remains after mitigations".

Along with these definitions, it is possible to organize the different types of risks in different classes and their possible combinations. In Table 1 a first example of classification is shown, based on two characteristics that relate the origin and generation of the risk (organizational perimeter) with the possibilities of intervention (controllability of risk).

		Controllability		
		Controllable	Partially controllable	Uncontrollable
Organization	Internal	Quality and cost of products	Environmental impacts	Incidents and accidents
	External	Technological progress	Demand variation	Natural disasters

Table 1. Example of risk classification by perimeter

Further classifications can also be taken from the already mentioned risk management models, where the descriptive categories are represented as a function of different objectives and decision-making levels (Table 2).

Model	Dimension	Classes
Risk Management Standard [3]	Level of interaction (internal and external)	- Strategic risks (partner and market) - Financial risks (economic-financial cycle) - Operational risks (process) - Potential risks (social and territorial environment)
Strategy Survival Guide [7]	Decisional level	- External risks (PESTLE - Political, Economic, Socio-cultural, Technological, Legal/regulatory, Environmental) - Operational risks (delivery, capacity and capability, performance) - Change risks (change programs, new projects, new policies)
FIRM Risk Scorecard [8]	Area of impact	- Infrastructural risks - Financial risks - Market risks - Reputational risks
Enterprise Risk Management [4]	Area of impact	- Strategic risks - Operational risks - Reporting risks - Compliance risks

Table 2. Example of risk classification by target

Developing the classification to an extended level and considering all the sources of uncertainty that affects business targets, vulnerability of organizations can be assessed on five different areas (Table 3).

Risk Category	Risk factors
Demand (Customers)	- Number and size of customers
	- Changes in number and frequency of orders
	- Changes to orders
	- Seasonal and promotional effects
	- Forecasting
	- Warehouses and inventory
	- Level of innovation and competition
	- Life cycle of the product
	- Timing and mode of payment
	- Retention rate
Offer (Suppliers)	- Number and size of suppliers
	- Level of quality and performance
	- Level of flexibility and elasticity
	- Duration and variability of lead time
	- Length and mode of transfers
	- Forecasting and planning
	- Just-in-Time or Lean approaches
	- Cost efficiency
	- Price levels
	- Outsourcing
	- Internationalization
	- Disruption
Processes (Organization)	- Flexibility of production-distribution systems
	- Variability in process management
	- Variability in process performance
	- Level of productivity
	- Capacity
	- Handling
	- Operational and functional failures
	- Redundancy of backup systems (quantity and quality)
	- Profit margins
	- Technological standards
	- Technological innovation of product and process
	- Product customization
Network and collaboration (Relations)	- Trust and interdependence among partners
	- Level of collaboration
	- Design and development of relations
	- Level of integration
	- Level of service
	- Opportunism and information asymmetry in transactions
	- Bargaining power

Risk Category	Risk factors
	- Strategic objectives and mission
	- Corporate cultures
	- Business Logic
	- Relationship and stakeholder engagement
	- Social and administrative responsibility
	- Availability and reliability of information systems
	- Intellectual property
Environment (Externalities)	- Regulations
	- Policies
	- Laws
	- Taxes
	- Currency
	- Strikes
	- Natural events
	- Social events (i.e. terrorism)

Table 3. Risk classification by organization

4. Enterprise risk management for strategic planning

The competitiveness of an organization depends on its ability to create value for its stakeholders. The management maximizes the value when objectives and strategies are formulated in order to achieve an optimal balance between growth, profitability and associated risks, using resources in an efficient and effective way. These statements are the basic philosophy of "risk management business". As seen, all businesses face uncertain events and the challenge of management is to determine the amount of uncertainty acceptable to create value. The uncertainty is both a risk and an opportunity and can potentially reduce or increase the value of the company.

The Enterprise Risk Management (ERM) is the set of processes that deals with the risks and opportunities that have an impact on the creation or preservation of value. ERM is put in place by the Board of Administration, the management and other professionals in an organization to formulate strategies designed to identify potential events that may affect the business, to manage risk within the limits of acceptable risk and to provide reasonable assurance regarding the achievement of business targets. It is an ongoing and pervasive process that involves the whole organization, acted by people of different roles at all levels and throughout the corporate structure, both on its specific assets and on the company as a whole.

This definition is intentionally broad and includes key concepts, critical to understand how companies must manage risk, and provides the basic criteria to apply in all organizations, whatever their nature. The ERM enables to effectively deal with uncertainty, enhancing the company's ability to generate value through the following actions:

- *the alignment of strategy at acceptable risk:* management establishes the level of acceptable risks in evaluating strategies, setting objectives and developing mechanisms to manage the associated risks;

- *the improvement of the response to identified risks:* ERM needs a rigorous methodology to identify and select the most appropriate among several alternatives of responses to risks (avoid, reduce, share, accept the risk);

- *the reduction of contingencies and resulting losses:* companies, increasing their ability to identify potential events, assess the risks and formulate responses, reducing the frequency of unexpected events as well as the subsequent costs and losses.

- *the identification and management of multiple and correlated risks:* every business needs to face an high number of risks affecting different areas and the ERM facilitates the formulation of a unique response to clusters of risks and associated impacts;

- *the identification of opportunities:* through the analysis of all possible events, management is able to proactively identify and seize the opportunities that emerge;

- *the improvement of capital expenditure:* the acquisition of reliable information on risks allows management to effectively assess the overall financial needs, improving the allocation of resources.

These characteristics help management to achieve performance targets without wasting resources. Furthermore, it ensures the effectiveness of reporting in compliance with laws and regulations, so to prevent damages to corporate reputation and relative consequences. Summarizing, the ERM supports organizations to accomplish their goals while avoiding pitfalls and unexpected path.

5. The risk management process

The risk management process consists of a series of logical steps for analyzing, in a systematic way, the hazards, the dangers and the associated risks that may arise in the management of an organization. The goal is realized in giving maximum sustainable value to any activity, through a continuous and gradual process that moves from the definition of a strategy along its implementation. By understanding all potential positive and negative factors that affect the system, it is possible to increase the probability of success and reduce the level of uncertainty.

In particular, risk management protects and supports the requirements of the organization in its relationship with stakeholders through:

- a methodological framework that allows a consistent and controlled development of activities;

- the improvement of the decision-making process, creating priorities by really understanding the natural and exceptional variability of activities and their positive or negative effects;

- the contribution to a more efficient use and allocation of resources;

• the protection and enhancement of corporate assets and image;

• the development and support to the people and to their knowledge base.

Figure 2 represents a process of risk management in its different stages of development that are detailed in the following sections.

5.1. Risk assessment

Risk assessment is a sequence of various activities aimed at identifying and evaluate the set of risks that the organization has to face. The international literature offers several techniques of modeling and decision-making [9-10] that can become part of the analysis.

The results of risk assessment can be summed up in two outputs that address the following stages of treatment and control:

• the risk profile;

• the risk appetite.

The risk profile represents the level of overall exposure of the organization, defining in a complete way the complexity of the risks to be managed and their ranking, according to their entity and significance. A segmentation for entities (areas, functions, people, sites) or decisional levels and the actual measures of treatment and control complete the profile. This takes to the expression of the:

• *gross profile:* the level of exposure to the events without any measure of treatment;

• *net profile:* the level of exposure, according to the measures of treatment in place (if effective or not);

• *net future profile:* the level of exposure surveyed after all the measures of treatment are implemented.

The definition of the risk appetite is a key outcome of the assessment process: on the one hand it is appropriate to draft it before the risk identification (where the level of accuracy of analysis can also depend on the risk appetite itself), on the other it is absolutely necessary to fix it before taking any decision about the treatment.

In any case, the risk appetite presents two different dimensions according to the scope of analysis:

• *threat:* the threshold level of exposure considered acceptable by the organization and justifiable in terms of costs or other performance;

• *opportunity:* what the organization is willing to risk to achieve the benefits in analysis, compared with all the losses eventually arising from a failure.

The so defined risk appetite can be adjusted through the delegation of responsibilities, strengthening the capability of taking decisions at different levels according to cost dynamics.

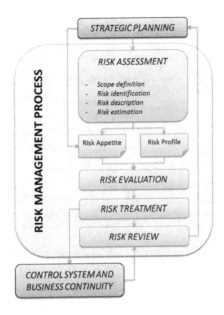

Figure 2. Risk Management process

5.1.1. Scope definition

The target of this stage is the identification of assets and people exposed to the risks and the identification of factors that determine the risks themselves. The definition of the scope has a critical importance in order to evaluate internal and external influences on the organization.

As this analysis requires requires a thorough knowledge of the environmental components (business, market, political, social and cultural issues), it has to be developed for all the decision-making levels (strategic, tactical and operational) and for all the stakeholders. Furthermore, the relationships with the output of the strategic planning have to be determined as the relevance of a risk and the priorities of interventions can be identified only with reference to the targets to uncertainty, while the eventual impact can vary widely according to a proper assignment and commitment of resources.

Despite this stage is found to be of fundamental importance for the effectiveness of the others, in particular for the identification of risks, it is too often executed with an inappropriate level of attention or it is not developed at all.

5.1.2. Risk identification

The identification of risks allows to acquire knowledge on possible future events, trying to measure the level of exposure of the organization. The target is to identify all the significant source of uncertainty in order to describe and proactively manage different scenarios. The identification is the first step to define the *risk profile* and the *risk appetite* of the organization.

This activity has to be repeated continuously and can be divided into two distinct stages:

- initial identification of risks: to be developed for organizations without a systematic approach to risk management. It is required to gather information on hazards and their possible evolutions;

- ongoing identification of risks: to update the risk profile of an organization and its relations, taking into account the generation of new risks or modifications to the already identified ones.

All the process mapping techniques are extremely useful to associate and connect risks with activities (Figure 3). The level of detail is determined by the necessity of identifying the specific impact associated with risks, of assigning responsibility of management and defining the subsequent actions to ensure control.

This can be developed with the support of external consultants or through a self-assessment which, if conducted with adequate methodological tools, provides a better awareness of the profile and an upgrade of the management system.

Among the others, the most common and widely used (successfully tested in other fields as for marketing and quality management) are:

- techniques of data collection and statistical analysis;

- techniques of problem finding and problem solving;

- SWOT analysis and Field Force;

- benchmarking with competitors or best in class.

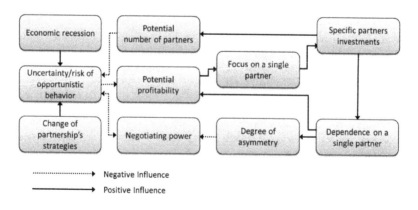

Figure 3. Example of risk identification for collaboration risks

5.1.3. Risk description

The results of identification should be developed in an appropriate stage of description by means of specific information support systems (i.e. Risk Register, table 4). Depending on the scope, the documentary support can assume different forms to improve the sharing of information and increase efficiency of management. Whatever the solution adopted for the description, this has to be dynamically completed with data coming from the different stages of the risk management process and updated according to changes of internal and external context. Inheriting the best practices already in use for environmental and safety management systems, when the risks are in any way related to regulations (i.e. Sarban Oaxley's act), a Compliance Register has to be associated to the Risk Register to ensure the conformity to requirements.

5.1.4. Risk estimation

The risk assessment has to end up with the association of a qualitative or quantitative measure of any risk, in terms of technical, economic or financial intensity. The choice of the methodology is related to the level of details required by the comparison among risk profile and risk appetite and to the availability of data and information. The metrics can refer to:

- probability of occurrence and magnitude of the effects and impacts;

- value at risk or vulnerability, which is the possible value of benefits or threats in relation to the characteristics of the organization.

The estimation of risk can be performed using different qualitative, quantitative or mixed criteria each with a different level of detail and reliability of the results. While the first are characterized by a strong subjectivity that only a high level of experience can compensate, the second need harmonization and conversion of the scales and of the values found. The choice is also related to the desired output of the stage, typically a hierarchical ordering of the risks identified (e.g. some types of exposures and tolerability are defined by regulations, especially for safety and environment). Examples of simple evaluation criteria, according to the already mentioned reference model, are shown in table 4, 5 and 6.

Identification code	ID to associate and create links among information
Category	According to the classification adopted
Organizational level	Corporate, business unit, site, process or activities involved
Related target	Relation to the strategic planning and decisional level
Stakeholders	Involvement of the different stakeholders
Regulation	Relation to compulsory (laws or directives) or voluntary (procedures) requirements
Description	Extended description of the event and its possible evolutions (hazard)
Causes	First, second and third level causes (direct or indirect)

Consequences	Description of impacts (direct or indirect)
Emergency	Potential emergency related to the risk and associate plans of recovery
Inherent risk	Combination of the probability (or frequency) of the event and the impact or relevance of the effects
Risk appetite	Threshold level of tolerance of the specific risk
Treatment	Extended description of the mitigations
Residual risk	Estimation of the risk after the of mitigation
Control	Extended description of the control
Risk owner	Responsibility of the risk and related activities
Control owner	Responsibility of the control and related activities

Table 4. Risk register

High	- financial impact on the organization probably higher than xxx € - notable impact on strategies or operations of the organization - notable involvement of the stakeholders
Medium	- financial impact on the organization probably among yyy € and xxx € - reasonable impact on strategies or operations of the organization - reasonable involvement of the stakeholders
Low	- financial impact on the organization probably lower than yyy € - limited impact on strategies or operations of the organization - limited involvement of the stakeholders

Table 5. Impacts of threats and opportunities [3]

Value	Indicator	Description
High (Probable)	Probable every year or in more than 25% of cases	- possible happening of the event in the period of analysis, with many repetitions - it happened recently
Medium (Possible)	Probable in 10 years or in less than 25% of cases	- possible happening of the event in the period of analysis, with some repetitions - difficulties in forecasting and controllability - data on past events exist
Low (Remote)	Improbable in 10 years or in less than 2% of cases	- mostly likely it never happens - it never happened

Table 6. Probability of the event: threats [3]

Value	Indicator	Description
High (Probable)	Probable advantages in the year or in more than 75% of cases	- clear opportunity with reasonable certainty - to act in the short period with the actual processes
Medium (Possible)	Reasonable advantages in the year or between 75% and 25% of cases	- achievable opportunity that requires an accurate management - opportunity beyond the programs
Low (Remote)	Possible advantages in the midterm or in less than 25% of cases	- possible opportunity that has to be deeply examined - opportunity with low probability of success according to the actual resources involved

Table 7. Probability of the event: opportunities [3]

5.2. Risk evaluation

The evaluation of risks provides a judgment concerning the acceptability or the need of mitigations, according to the comparison between the risk profile and the risk appetite. The stage is a decision-making process in which, if the risk is acceptable, the assessment can be terminated, otherwise it goes on to next stage of treatment and management. To verify the acceptability after the interventions, the results of the mitigations have to be iteratively compared to the expected targets. At this stage it is possible to use, with adaptation when necessary, methods and techniques widely tested in safety management:

- *Event Tree Analysis (ETA) and Fault Tree Analysis (FTA)*: analysis of the cause-effect tree of the risk profile. The top event (an event that is at the end of the shaft) is usually a cause of loss of value in the organization, related to exclusionary or concurrent events of a lower-level type;

- *Failure Modes Effects Analysis (FMEA) and Failure Modes Effects and Criticality Analysis (FMECA)*: FMEA is a technique that allows a qualitative analysis of a system, decomposing the problem in a hierarchy of functions up to a determined level of detail. For each of the constituents, possible "failure modes" (adverse events) are identified and actions to eliminate or reduce the effects can be considered. FMECA adds a quantitative assessment of the criticalities: for each mode, an index is calculated as the combination of the occurrence of the event, the severity of its effects and the detectability of the symptoms;

- *Hazard and Operability (HAZOP) analysis*: qualitative methodology that has both deductive (search for causes) and inductive (consequence analysis) aspects. The method seeks for the risks and operational problems that degrade system performances and then find solutions to the problems identified;

- *Multi-criteria decision tools (i.e. Analytic Hierarchy Process and Analytic Network Process)*: decision support techniques for solving complex problems in which both qualitative and quantitative aspects have to be considered. Through a hierarchical or network modeling, the definition of a ranking of the critical aspects of the problem is enabled. Multi-criteria decision tools give an effective support mainly where the consequences of an event can be both positive and negative, applying cost-benefit analysis.

5.3. Risk treatment

Treatment of risks must be determined after a first evaluation and comparison of the risk profile and the risk appetite of the organization. The actions arising from this decision-making stage can be classified according to the following scheme:

- *terminate*: remove, dispose or outsource, where possible, the factors that can cause the risk. It can take the organization to refuse opportunities if the value at risk is higher than the risk appetite;

- *treat*: develop measures of mitigation in order to intervene on the values of significance of the risk, reducing the probability of occurrence (prevention), the potential impacts of the effects (protection) or determining actions of restoring (recovery) after damages are occurred. Passing from prevention to protection and recovery, the capability of controlling risks tends to decrease, while increasing the exposure of the organization;

- *tolerate*: accept the risk profile as compatible with the risk appetite, in relation to the resource involved;

- *transfer*: transfer the impacts to third parties through, for example, insurances or risk sharing actions. Possible uncertain effects are converted in certain payments;

- *neutralize*: balance two or more risk, for example increasing the number of unit exposed, so that they can cancel each other;

- *take the opportunity*: when developing actions of treatment, opportunities of positive impacts can be identified and explored.

5.4. Risk review

The key target of the review stage is to monitor the changes in the risk profile and in the risk appetite of the organization and to provide assurance to all stakeholders that the risk management process is appropriate to the context, effectively and efficiently implemented.

The frequency of the review should be determined depending on the characteristics of the risk management system, to execute:

- *a review of the risks*, to verify the evolution of already existing risks and the arise of new risks, assessing their entity;

- *a review of the risk management process*, to ensure that all activities are under control and to detect changes in the structure of the process.

6. The control system

The conceptual path that characterizes this approach to risk management is strictly related to the existence of an indissoluble connection between risks and controls. Most current control systems recognize the risk as part of the corporate governance that has to be:

• continuous, integrating control in the decision-making processes;

• pervasive, spreading the risk management at all decisional levels;

• formalized, through the use of clear and shared methodologies;

• structured, through the adoption of suitable organizational solutions.

The control system traditionally represents a reactive approach in response to adverse events, fragmented in different areas and occasional frequencies. From a standard dimension, generally limited to financial risks or internal audit, it has to evolve towards a proactive continuous process, results-oriented and with widespread responsibility. The challenge for management is to determine a sustainable amount of uncertainty to create value in relation to the resources assigned, facing a costs and benefits trade-off where the marginal cost of control is not greater than the benefit obtained.

The main components of the control system can be summarized as follows:

• *control environment*: it is the base of the whole system of controls as it determines the sensitivity level of management and staff on the execution of processes. The adoption and dissemination of codes of ethics and values, policies and management style, the definition of a clear organizational structure and responsibilities (including specific bodies of internal control), the development of professional skills of human resources are the elements that constitute this environment;

• *control activities*: it is the operational component of the control system, configured as a set of initiatives and procedures to be executed, both on process and interfaces, to reduce business risks to a reasonable level, ensuring the achievement of the targets;

• information and communication: a structured information system at all levels enables the control on processes, recomposing flows managed by different subsystems and applications that need to be integrated. Adequate information, synthetic and timely, must be provided to allow the execution of activities, taking responsibilities and ensuring monitoring;

• *monitoring*: it is the continuous supervision and periodic evaluation of the performances of the control system. The scope and techniques of monitoring depend on the results of the risk assessment and on the effectiveness of the procedures in order to ensure that the controls are in place to efficiently reduce the risks.

7. The business continuity

But how can organizations deal with those types of risks generally unknown and not predict-able? The answer comes from a different kind of strategic vision that is not only based on the analysis of identified risks but looks at the possible modes of disruption of processes regardless of the cause. For example, once defined the logistics distribution as a key factor for the success of the business, you can evaluate how to recover any link regardless of the specific reasons of interruption.

The Business Continuity Management is an approach generally used in case of possible serious consequences related to crisis or emergency [11-13]: an organization that evaluates the effects of damage to a warehouse caused by a sudden storm or defines actions following the failure of a partner is performing risk management; when it arranges structured actions related to the unavailability of a warehouse or a provider moves up to the level of Business Continuity Management and its main features:

- analysis and identification of the elements of the organization that may be subject to inter-ruption, unavailability and related effects;

- definition of action plans and programs to be implemented when an element is missing, to ensure the continuity of material and information flows or recover as quickly as possible;

- monitoring of processes to anticipate possible crises or to start emergency plans;

- establishment of systematic test of recovery plans;

- once recovered, structured analysis of events to evaluate the success of the initiatives, the efficiency of the plans and their revision.

The Business Continuity Management accompanies organizations during disaster recovery of unexpected risks, particularly rare and with high magnitudes of the effects, where the opera-tions must be carried out with the utmost speed and effectiveness. Events such as the earth-quake in Kobe (Japan) in 1995, that caused more than 6,400 deaths, 100,000 demolished buildings, closed the major ports of the country for two months, having a general impact on industries for more than 100 billion dollars, can easily be presented for examples of disasters. At the same time, also much smaller events can be recognized as a disaster for small and medium-sized enterprises, such as the loss of a key customer, a huge credit not collected, a wrong industrial initiative, a failure of the production system or the breakdown of a relation-ship with a partner.on dollars, the other much smaller events also can be recognized as a disaster for small and medium-sized enterprises, such as the loss of a key customer, a huge credit not collected, a wrong industrial initiative, a system failure or breakdown of a relation-ship with a partner. In the same way, any loss related to a failure of an infrastructure can generate adverse effects as well as an incorrect definition of strategic processes or the indis-criminate and uncoordinated introduction of new methods such as just-in-time: the majority of negative events come from managerial mistakes that could be avoided rather than from the effects of real and unexpected emergencies.

A recovery plan must therefore meet the following requirements:

- ensure the physical integrity of employees, customers, visitors and in any case all subjects interacting with current activities;

- protect as much as possible facilities and resources to ensure a rapid recovery;

- implement procedures to restore a minimum level of service, while reducing the impact on the organization;

- work with partners to redefine the appropriate services: once reorganized the internal activities, it is necessary to seek outside to assess the effects and consequences of actions taken;

- return all processes to performance standard in time and at reasonable cost: the speed with which the repairs must be carried out is balanced with the associated costs.

8. Conclusions

The main advantages that companies could obtain from Enterprise Risk Management were deeply investigated in the sections above. Anyway, this novel approach could present some difficulties, common to many businesses, related to the absence of a culture of strategic planning aimed at prevention rather than response, to a general lack of professionals and of appropriate tools capable to really integrate processes. But while complexity is becoming a part of the corporate governance system, absorbing a great amount of time and resources, the need for competitiveness requires a specific attention to performances and results. A new attitude of organizations towards risk-sensitive areas, able to ensure the coordination among all its components, helps to transform the management of risk from a cost factor to an added value. This business view, allows, with a little effort, to reduce the overall risk of the company and helps the dialogue among business functions and with the stakeholders.

Author details

Giulio Di Gravio*, Francesco Costantino and Massimo Tronci

*Address all correspondence to: giulio.digravio@uniroma1.it

Department of Mechanical and Aerospace Engineering, University of Rome "La Sapienza", Italy

References

[1] Minahan T.A. The Supply Risk Benchmark Report. Aberdeen Group; 2005.

[2] UK HM Treasury. Orange book management of risk – principles and concepts. 2004.

[3] Association of Insurance and Risk Managers (AIRMIC), GB Institute of Risk Manage-
     ment (IRM), ALARM National Forum for Risk Management. A Risk Management
     Standard. 2002

[4] AU Committee of Sponsoring Organizations of the Treadway Commission. Enterprise
     Risk Management – Integrated Framework. 2004.

[5] European Foundation for Quality Management. EFQM framework for risk manage-
     ment. EFQM; 2005.

[6] ISO Guide 73:2009 - Risk Management - Vocabulary. ISO; 2009

[7] UK Prime Minister's Strategy Unit. Strategy survival guide. 2004.

[8] Information Security Forum. Fundamental Information Risk Management (FIRM). ISF;
     2000.

[9] ISO 31000:2009. Risk management - Principles and guidelines. ISO; 2009

[10] ISO/IEC 31010:2009. Risk Management - Risk Assessment Techniques. ISO; 2009

[11] British Standard Institute. PAS56: Guide to Business Continuity Management. BSI;
     2006.

[12] Chartered Management Institute. Business Continuity Management. CMI; 2005.

[13] Department of Trade and Industry. Information Security: Understanding Business
     Continuity Management. Stationery Office; 2006.

Chapter 9

An Overview of Human Reliability
Analysis Techniques in Manufacturing Operations

Valentina Di Pasquale, Raffaele Iannone,
Salvatore Miranda and Stefano Riemma

Additional information is available at the end of the chapter

1. Introduction

In recent years, there has been a decrease in accidents due to technical failures through technological developments of redundancy and protection, which have made systems more reliable. However, it is not possible to talk about system reliability without addressing the failure rate of all its components; among these components, "man" – because his rate of error changes the rate of failure of components with which he interacts. It is clear that the contribution of the human factor in the dynamics of accidents – both statistically and in terms of severity of consequences – is high [2].

Although valid values are difficult to obtain, estimates agree that errors committed by man are responsible for 60–90% of the accidents; the remainder of accidents are attributable to technical deficiencies [2,3,4]. The incidents are, of course, the most obvious human errors in industrial systems, but minor faults can seriously reduce the operations performances, in terms of productivity and efficiency. In fact, human error has a direct impact on productivity because errors affect the rates of rejection of the product, thereby increasing the cost of production and possibly reduce subsequent sales. Therefore, there is need to assess human reliability to reduce the likely causes of errors [1].

The starting point of this work was to study the framework of today's methods of human reliability analysis (HRA): those quantitative of the first generation (as THERP and HCR), those qualitative of second (as CREAM and SPAR-H), and new dynamic HRA methods and recent improvements of individual phases of HRA approaches. These methods have, in fact, the purpose of assessing the likelihood of human error – in industrial systems, for a given operation, in a certain interval of time and in a particular context – on the basis of models that describe, in a more or less simplistic way, the complex mechanism that lies behind the single human action that is potentially subject to error [1].

The concern in safety and reliability analyses is whether an operator is likely to make an incorrect action and which type of action is most likely [5]. The goals defined by Swain and Guttmann (1983) in discussing the THERP approach, one of the first HRA methods developed, are still valid: The objective of a human reliability analysis is 'to evaluate the operator's contribution to system reliability' and, more precisely, 'to predict human error rates and to evaluate the degradation to human–machine systems likely to be caused by human errors in association with equipment functioning, operational procedures and practices, and other system and human characteristics which influence the system behavior' [7].

The different HRA methods analysed allowed us to identify guidelines for determining the likelihood of human error and the assessment of contextual factors. The first step is to identify a probability of human error for the operation to be performed, while the second consists of the evaluation through appropriate multipliers, the impact of environmental, and the behavioural factors of this probability [1]. The most important objective of the work will be to provide a simulation module for the evaluation of human reliability that must be able to be used in a dual manner [1]:

• In the preventive phase, as an analysis of the possible situation that may occur and as evaluation of the percentage of pieces discarded by the effect of human error;

• In post-production, to understand what are the factors that influence human performance so they can reduce errors.

The tool will also provide for the possibility of determining the optimal configuration of breaks through use of a methodology that, with assessments of an economic nature, allow identification of conditions that, in turn, is required for the suspension of work for psychophysical recovery of the operator and then for the restoration of acceptable values of reliability [1].

2. Literature review of HRA methods

Evidence in the literature shows that human actions are a source of vulnerability for industrial systems, giving rise to HRA that aims to deepen the examination of the human factor in the workplace [1]. HRA is concerned with identifying, modelling, and quantifying the probability of human errors [3]. Nominal human error probability (HEP) is calculated on the basis of operator's activities and, to obtain a quantitative estimate of HEP, many HRA methods utilise performance shaping factors (PSF), which characterise significant facets of human error and provide a numerical basis for modifying nominal HEP levels [24]. The PSF are environmental factors, personal, or directed to activities that have the potential to affect performance positively or negatively; therefore, identifying and quantifying the effects of a PSF are key steps in the process of HRA [3]. Another key step concerns interpretation and simulation of human behaviour, which is a dynamic process driven by cognitive and behavioural rules, and influenced by physical and psychological factors. Human behaviour, although analysed in numerous studies, remains difficult to fully represent in describing all the nuances that distinguish it [1]. It is abundantly clear how complex an effort has been made in the literature to propose models of human behaviour, favoring numerical values of probability of error to predict and prevent unsafe behaviours. For this reason, the study of human reliability can be seen as a specialised scientific subfield – a hybrid between psychology, ergonomics, engineering, reliability analysis, and system analysis [4].

The birth of HRA methods dates from the year 1960, but most techniques for assessment of the human factor, in terms of propensity to fail, have been developed since the mid-'80s. HRA techniques or approaches can be divided essentially into two categories: first and second generation. Currently, we come to HRA dynamic and methods of the third generation, understood as an evolution of previous generations.

2.1. First generation HRA methods

The first generation HRA methods have been strongly influenced by the viewpoint of probabilistic safety assessment (PSA) and have identified man as a mechanical component, thus losing all aspects of dynamic interaction with the working environment, both as a physical environment and as a social environment [33]. In many of these methods – such as Technique for Human Error Rate Prediction (THERP) [2, 3, 13–15], Accident Sequence Evaluation Program (ASEP) [16], and Human Cognition Reliability (HCR) [2] – the basic assumption is that because humans have natural deficiencies, humans logically fail to perform tasks, just as do mechanical or electrical components. Thus, HEP can be assigned based on the characteristics of the operator's task and then modified by performance shaping factors (PSF). In the first HRA generation, the characteristics of a task, represented by HEPs, are regarded as major factors; the context, which is represented by PSFs, is considered a minor factor in estimating the probability of human failure [8]. This generation concentrated towards quantification, in terms of success/failure of the action, with less attention to the depth of the causes and reasons of human behaviour, borrowed from the behavioural sciences [1].

THERP and approaches developed in parallel – as HCR, developed by Hannaman, Spurgin, and Lukic in 1985 – describe the cognitive aspects of operator's performance with cognitive modelling of human behaviour, known as model skill-rule-knowledge (SKR) by Rasmussen (1984) [2]. This model is based on classification of human behaviour divided into skill-based, rule-based, and knowledge-based, compared to the cognitive level used (see Fig. 1).

The attention and conscious thought that an individual gives to activities taking place decreases moving from the third to first level. This behaviour model fits very well with the theory of the human error in Reason (1990), according to which there are several types of errors, depending on which result from actions implemented according to the intentions or less [2]. Reason distinguishes between: slips, intended as execution errors that occur at the level of skill; lapses, that is, errors in execution caused by a failure of memory; and mistakes, errors committed during the practical implementation of the action. In THERP, instead, wrong actions are divided into errors of *omission* and errors of *commission*, which represent, respectively, the lack of realisation of operations required to achieve the result and the execution of an operation, not related to that request, which prevents the obtainment of the result [1, 4].

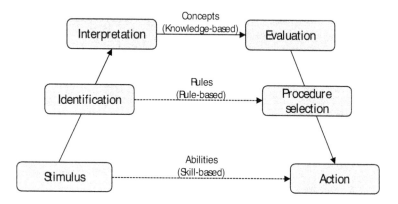

Figure 1. Rasmussen's SKR model [2].

The main characteristics of the methods can be summarised as follows [9]:

• Binary representation of human actions (success/failure);

• Attention on the phenomenology of human action;

• Low concentration on human cognitive actions (lack of a cognitive model);

• Emphasis on quantifying the likelihood of incorrect performance of human actions;

• Dichotomy between errors of omission and commission;

• Indirect treatment of context.

Among the first generation techniques are: absolute probability judgement (APJ), human error assessment and reduction technique (HEART), justified human error data information (JHEDI), probabilistic human reliability analysis (PHRA), operator action tree system (OATS), and success likelihood index method (SLIM) [31,32]. Among these, the most popular and effectively method used is THERP, characterised as other first generation approaches by an accurate mathematical treatment of the probability and error rates, as well as computer programs well-structured for interfacing with the trees for evaluation of human error of a fault event and trees [11]. The base of THERP is event tree modelling, where each limb represents a combination of human activities, influences upon these activities, and results of these activities [3]. The basic analytical tool for the analysis of human reliability is represented with the graphics and symbols in Figure 2.

First generation HRA methods are demonstrated with experience and use, not able to provide sufficient prevention and adequately perform its duties [10]. The criticism of base to the adequacy of the traditional methods is that these approaches have a tendency to be descriptive of events in which only the formal aspects of external behaviour are observed and studied in terms of errors, without considering reasons and mechanisms that made them level of cognition. These methods ignore the cognitive processes that underlie human performance and, in fact, possess a cognitive model without adequate human and psychological realism. They are often criticised for not having considered the impact of factors such as environment, organisational factors, and other relevant PSFs; errors of commission; and for not using proper methods of judging experts [4,10,25]. Swain remarked that "all of the above HRA inadequacies often lead to HRA analysts assessing deliberately higher estimates of HEPs and greater uncertainty bounds, to compensate, at least in part, for these problems" [4]. This is clearly not a desirable solution.

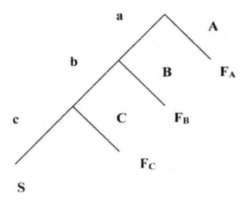

Figure 2. Scheme for the construction of a HRA-THERP event tree [2]: Each node in the tree is related to an action, the sequence of which is shown from the top downwards. Originating from each node are two branches: The branch to the left, marked with a lowercase letter, indicates the success; the other, to the right and marked with the capital letter, indicates the failure.

Despite the criticisms and inefficiencies of some first-generation methods, such as THERP and HCR, they are regularly used in many industrial fields, thanks to their ease of use and highly quantitative aspects.

2.2. Second generation HRA methods

In the early 1990s, the need to improve HRA approaches interested a number of important research and development activities around the world. These efforts led to much progress in first generation methods and the birth of new techniques, identified as second generation. These HRA methods have been immediately unclear and uncertain, substantially because the methods have been defined in terms of what should not be – that is, they should be as the first generation of HRA methods [5]. While the first generation HRA methods are mostly behavioural approaches, the second generation HRA methods aspire to be of conceptual type [26]. The separation between generations is evident in the abandonment of the quantitative approach of PRA/PSA in favour of a greater attention to qualitative assessment of human error. The focus shifted to the cognitive aspects of humans, the causes of errors rather than their frequency, the study of the interaction of the factors that increase the probability of error, and the interdependencies of the PSFs [1].

Second generation HRA methods are based on a cognitive model more appropriate to explain human behaviour. It is evident that any attempt at understanding human performance needs to include the role of human cognition, defined as "the act or process of knowing including both awareness and judgement" by an operator [1]. From the HRA practitioner's perspective, the immediate solution to take into consideration human cognition in HRA methods was to introduce a new category of error: "cognitive error", defined both as failure of an activity that is predominantly of a cognitive nature and as the inferred cause of an activity that fails [4]. For example, in CREAM, developed by Erik Hollnagel in 1993, maintained division between logical causes and consequences of human error [5]. The causes of misbehaviour (genotypes) are the reasons that determine the occurrence of certain behaviours, and the effects (phenotypes) are represented by the incorrect forms of cognitive process and inappropriate actions [2,17,25].

Moreover, the second generation HRA methods have aimed at the qualitative assessment of the operator's behaviour and the search for models that describe the interaction with the production process. Cognitive models have been developed, which represent the process logical–rational of the operator and summarise the dependence on personal factors (such as stress, incompetence, etc.) and by the current situation (normal conduction system, abnormal conditions, or even emergency conditions), and models of man–machine interface, which reflect the control system of the production process [33]. In this perspective, man must be seen in an integrated system, men–technology–organisation (MTO), or as a team of operators (men) who collaborate to achieve the same objective, intervening in the mechanical process (technology) within a system of organisation and management of the company (organisation) and, together, represent the resources available [1,6].

The CREAM operator model is more significant and less simplistic than that of first generation approaches. The cognitive model used is the contextual control model (COCOM), based on the assumption that human behaviour is governed by two basic principles: the cyclical nature of human cognition and the dependence of cognitive processes from context and working environment. The model refers to the IPS paradigm and considers separately the cognitive functions (perception, interpretation, planning and action) and their connection mechanisms and cognitive processes that govern the evolution [2,4,5,8]. The standardised plant analysis risk–human reliability analysis method (SPAR-H) [11,12,34] is built on an explicit information-processing model of human performance, derived from the behavioural sciences literature. An information-processing model is a representation of perception and perceptual elements, memory, sensory storage, working memory, search strategy, long-term memory, and decision-making [34]. The components of the behavioural model of SPAR-H are presented in Figure 3.

A further difference between generations relates to the choice and use of PSF. None of the first generation HRA approaches tries to explain how PSFs exert their effect on performance; moreover, PSFs – such as managerial methods and attitudes, organisational factors, cultural differences, and irrational behaviour – are not adequately treated in these methods. PSFs in the first generation were mainly derived by focusing on the environmental impacts on operators, whereas PSFs in the second generation were derived by focusing on the cognitive impacts on operators [18]. The PSFs of both generations were reviewed and collected in a single taxonomy of performance influencing factors for HRA [16].

Human Behavior Model
Individual Factors

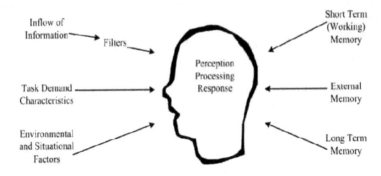

Figure 3. Model of human performance [12].

Among the methods of the second generation can be mentioned: a technique for human error analysis (ATHEANA), Cognitive Environmental Simulation (CES), Connectionism Assessment of Human Reliability (CAHR) and Méthode d'Evaluation de la Réalisation des Missions Opérateur pour la Sûreté (MERMOS) [31,32].

Many proposed second generation methods still lack sufficient theoretical or experimental bases for their key ingredients. Missing from all is a fully implemented model of the underlying causal mechanisms linking measurable PSFs or other characteristics of the context of operator response. The problem extends to the quantification side, where the majority of the proposed approaches still rely on implicit functions relating PSFs to probabilities [25]. In short, some of the key shortcomings that motivated the development of new methods still remain unfulfilled. Furthermore, unlike first generation methods, which have been largely validated [13–15], the second generation has yet to be empirically validated [32].

There are four main sources of deficiencies in current HRA methods [3]:

- Lack of empirical data for model development and validation;

- Lack of inclusion of human cognition (i.e. need for better human behaviour modelling);

- Large variability in implementation (the parameters for HRA strongly depend on the methodology used)

- Heavy reliance on expert judgement in selecting PSFs and use of these PSFs to obtain the HEP in human reliability analysis.

2.3. Last generation

In recent years, the limitations and shortcomings of the second generation HRA methods have led to further developments related to the improvement of pre-existing methods. The only method now defined as third generation is nuclear action reliability assessment (NARA) and is, in fact, an advanced version of HEART for the nuclear field. The shortcomings in the second generation, highlighted above, have been the starting point of HRA experts for new research and improvement of existing methods.

Some of the more recent studies have focused on lack of empirical data for development and validation of an HRA model and were intended to define the database HRA, which may provide the methodological tools needed to make greater use of more types of information in future HRAs and reduce uncertainties in the information used to conduct human reliability assessments. Currently, there are some databases for HRA analysts that contain the human error data with cited sources to improve the validity and reproducibility of HRA results. Examples of databases are the human event repository and analysis (HERA) [17] and the human factors information system (HFIS).

The PSFs are an integral part of the modelling and characterisation of errors and play an impor-tant role in the process of human reliability assessment; for this reason in recent years, HRA experts have focused their efforts on PSFs. Despite continuing advances in research and applications, one of the main weaknesses of current HRA methods is their limited ability to model the mutual influence among PSFs, intended both as a dependency among the states of the PSFs' dependen-cy among PSFs' influences (impacts) on human performance (Fig. 4) [20,26].

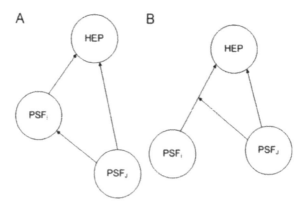

Figure 4. Possible types of dependency among PSFs: (A) dependency between the states (the presence) of the PSFs and (B) dependency between the state of the PSFj and the impact of PSFi over the HEP [20].

Some HRA methods – such as CREAM, SPAR-H, and IDAC – try to provide guidance on how to treat dependencies at the level of the factor assessments but do not consider that a PSF category might depend on itself and that the presence of a specific PSF might modulate the impact of another PSF on HEP; therefore, they do not adequately consider the relationships and dependencies between PSFs [20]. Instead, De Ambroggi and Trucco's (2011) study deals with the development of a framework for modelling the mutual influences existing among PSFs and a related method to assess the importance of each PSF in influencing performance of an operator, in a specific context, considering these interactions (see Fig. 5).

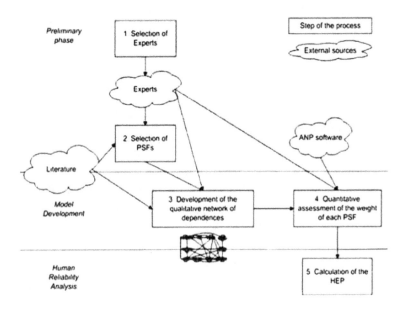

Figure 5. The procedure for modelling and evaluation of mutual influences among PSFs (De Ambroggi and Trucco 2011)

Another limitation of current HRA methods is the strong dependence on expert opinion to assign values to the PSFs; in fact, during this assignment process, subjectivity plays an important role, causing difficulties in assuring consistency. To overcome this problem and obtain a more precise estimation, Park and Lee (2008) suggest a new and simple method: AHP–SLIM [19].This method combines the decision-making tool AHP – a multicriteria decision method for complex problems in which both qualitative and quantitative aspects are considered to provide objective and realistic results – with success likelihood index method (SLIM), a simple, flexible method of the expert judgement for estimating HEPs [6,19]. Therefore through a type of HEP estimation using an analytic hierarchy process (AHP), it is possible to quantify the subjective judgement and confirm the consistency of collected data (see Fig. 6).

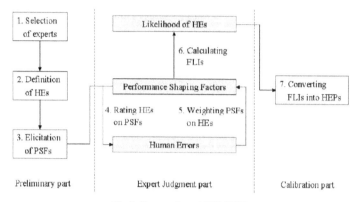

Fig. 2. The procedure of AHP–SLIM.

Figure 6. AHP–SLIM procedure scheme [19].

The real development concerns, however, are the so-called methods of reliability dynamics. Cacciabue [7] outlined the importance of simulation and modelling of human performance for the field of HRA. Specifically, simulation and modelling address the dynamic nature of human performance in a way not found in most HRA methods [23]. A cognitive simulation consists of the reproduction of a cognition model using a numerical application or computation [21,22].

As depicted in Figure 7, simulation and modelling may be used in three ways to capture and generate data that are meaningful to HRA [23]:

• The simulation runs produce logs, which may be analysed by experts and used to inform an estimate of the likelihood of human error;

• The simulation may be used to produce estimates PSFs, which can be quantified to produce human error probabilities (HEPs);

• A final approach is to set specific performance criteria by which the virtual performers in the simulation are able to succeed or fail at given tasks. Through iterations of the task that systematically explore the range of human performance, it is possible to arrive at a frequency of failure (or success). This number may be used as a frequentist approximation of an HEP.

Concurrent to the emergence of simulation and modelling, several authors (e.g. Jae and Park 1994; Sträter 2000) have posited the need for dynamic HRA and begun developing new HRA methods or modifying existing HRA methods to account for the dynamic progression of human behaviour leading up to and following human failure events (HFEs) [23]. There is still not a tool for modelling and simulation that fully or perfectly combines all the basic elements of simulation HRA. There is, however, a significant work in progress, as for the simulator PROCOS, developed by Trucco and Leva in 2006 or for the IDAC system, which combines a realistic plant simulator with a system of cognitive simulation capable of modelling the PSF. In addition to systems such as MIDAS, in which the modelling of the error was already present, further efforts are to instill the PSF of SPAR-H in the simulation system [24]. PROCOS [21,22] is a probabilistic cognitive simulator for HRA studies, developed to support the analysis of human reliability in operational contexts complex. The simulation model comprised two cognitive flow charts, reproducing the behaviour of a process industry operator. The aim is to integrate the quantification capabilities of HRA methods with a cognitive evaluation of the operator (see Fig. 8).

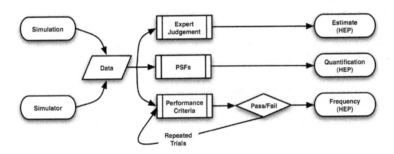

Figure 7. Uses of simulation and modelling in HRA [23].

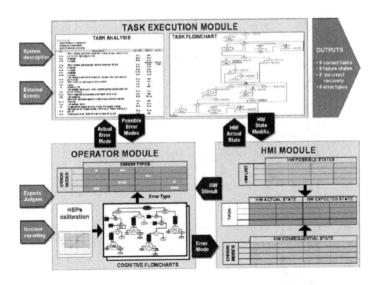

Figure 8. Architecture of PROCOS simulator [21].

The model used for the configuration of the flow diagram that represents the operators is based on a combination of PIPE and SHELL. The two combined models allow for representation of the main cognitive processes that an operator can carry out to perform an action (PIPE) and describe the interaction among procedures, equipment, environment and plants present in the working environment, and the operator, as well as taking into account the possibility of interaction of the operator with other operators or supervisors (SHELL).

The IDAC model [25–30] is an operator behaviour model developed based on many relevant findings from cognitive psychology, behavioural science, neuroscience, human factors, field observations, and various first and second generation HRA approaches. In modelling cognition, IDAC combines the effects of rational and emotional dimensions (within the limited scope of modelling the behaviour of operators in a constrained environment) through a small number of generic rules-of-behaviour that govern the dynamic responses of the operator. The model constrained behaviour, largely regulated through training, procedures, standardised work processed, and professional discipline. This significantly reduces the complexity of the problem, as compared to modelling general human response. IDAC covers the operator's various dynamic response phases, including situation assessment, diagnosis, and recovery actions in dealing with an abnormal situation. At a high level of abstraction, IDAC is composed of models of information processing (I), problem-solving and decision-making (D), and action execution (A) of a crew (C). Given incoming information, the crew model generated a probabilistic response, linking the context to the action through explicit causal chains. Due to the variety, quantity, and details of the input information, as well as the complexity of applying its internal rules, the IDAC model can only be presently implemented through a computer simulation (see Fig. 9).

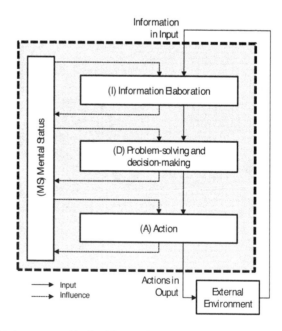

Figure 9. IDAC model of operator cognitive flow (Chang and Mosleh 2007).

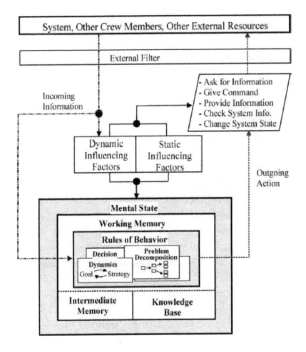

Figure 10. High-level vision of the IDAC dynamic response [25].

3. Literature review of rest breaks

One of the most important factors influencing the physical and mental condition of an employee – and, thus, his or her ability to cope with work – is the degree to which employees are able to recover from fatigue and stress at work. Recovery can be defined as the period of time that an individual needs to return to prestressor level of functioning following the termination of a stressor [35]. Jansen argued that fatigue should not be regarded as a discrete disorder but as a continuum ranging from mild, frequent complaints seen in the community to the severe, disabling fatigue characteristics of burnout, overstrain, or chronic fatigue syndrome [35]. It is necessary that recovery is properly positioned within this continuum not only in the form of lunch breaks, rest days, weekends or summer holidays, but even in the simple form of breaks or micro-pauses in work shifts.

Work breaks are generally defined as "planned or spontaneous suspension from work on a task that interrupts the flow of activity and continuity" [36]. Breaks can potentially be disruptive to the flow of work and the completion of a task. The potential negative consequences of breaks for the person being interrupted include loss of available time to complete a task, a temporary disengagement from the task, procrastination (i.e. excessive delays in starting or continuing work on a task), and the reduction in productivity; the break can lead to a loss of time to complete activities. However, breaks can serve multiple positive functions for the person being interrupted, such as stimulation for the individual performing a job that is routine or boring, opportunities to engage in activities that are essential to emotional wellbeing, job satisfaction, sustained productivity, and time for the subconscious to process complex problems that require creativity [36]. In addition, regular breaks seem to be an effective way to control the accumulation of risk during the industrial shift. The few studies on work breaks indicate that people need occasional changes... the shift or an oscillation between work and recreation, mainly when fatigued or working continuously for an extended period [36]. A series of laboratory studies in the workplace have been conducted to evaluate the effects of breaks in more recent times; however, there appears to be a single recent study that examined in depth the impact of rest breaks, focusing on the risk of injury. Tucker's study [37,38] focused attention on the risk of accidents in the workplace, noting that the inclusion of work breaks can reduce this risk. Tucker examined accidents in a car assembly plant, where workers were given a 15-minute break after each 2-hour period of continuous work. The number of accidents within each of four periods of 30 minutes between successive interruptions was calculated, and the risk in each period of 30 minutes was expressed in the first period of 30 minutes immediately after the break. The results are shown in Figure 5, and it is clear that the accident risk increased significantly, and less linearly, between the successive breaks. The results showed that rest breaks neutralise successfully accumulation of risk over 2 hours of continuous work. The risk immediately after a pause has been reduced to a rate close to that recorded at the start of the previous work period. However, the effects of the breaks are short-term recovery.

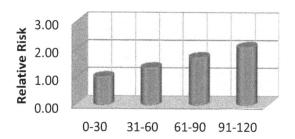

Minutes Since Last Break

Figure 11. The trend in relative risk between breaks [38].

A 2006 study by Folkard and Lombardi showed the impact of frequent pauses of different shift systems [39]. The results of these studies confirm that breaks, even for a short period of time, are positively reflected from physical and psychic viewpoints on the operator's work (see Fig. 12).

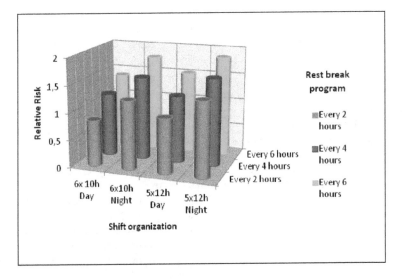

Figure 12. Effect of breaks in different shift systems [39].

Proper design of work–rest schedule that involves frequency, duration, and timing of rest breaks may be effective in improving workers' comfort, health, and productivity. But today, work breaks are not taken into proper consideration, and there are ongoing efforts to create systems that better manage the business in various areas, especially in manufacturing. From the analysis of the literature, in fact, there has been the almost total lack of systems for the management of work breaks in an automatic manner. The only exception is the software that stimulates workers at VDT to take frequent breaks and recommend performing exercises during breaks. The validity and effectiveness of this type of software has been demonstrated by several studies, including one by Van Den Heuvel [41] that evaluated the effects of work-related disorders of the neck and upper limbs and the productivity of computer workers stimulated to take regular breaks and perform physical exercises with the use of an adapted version of WorkPace, Niche Software Ltd., New, and that of McLean (2001) [40] that examined the benefits of micro-breaks to prevent onset or progression of cumulative trauma disorders for the computerised environment, mediated using the program Ergobreak 2.2.

In future, therefore, researchers should focus their efforts on the introduction of management systems of breaks and countering the rates of increase in the risk of accidents during long periods of continuous work to improve productivity.

4. Research perspectives in HRA

The previous paragraphs described the development of HRA methods from their origin to the last generation. In this generation, there are literally dozens of HRA methods from which to choose. However, many difficulties remain: Most of the techniques, in fact, do not have solid empirical bases and are essentially static, unable to capture the dynamics of an accident in progress or general human behaviour. Therefore, the limitations of current methods are natural starting point for future studies and work.

As described in this paper, the path has been paved for the next generation of HRA through simulation and modelling. The human performance simulation reveals important new data sources and possibilities for exploring human reliability, but there are significant challenges to be resolved, both as regards the dynamic nature of HRA versus the mostly static nature of conventional first and second generation HRA methods both for the weakness of the simulators themselves [23]. The simulator PROCOS, in particular, requires further optimisation, as evidenced by the same Trucco and Leva in [21]. Additionally, in its development, some sensitivity analysis has still to be performed on the main elements on which the simulator is based – blocks of the flow chart, decision block criteria, PSF importance – to test the robustness of the method [21]. Mosleh and Chang, instead, are conducting their studies to eliminate the weak points of IDAC as outlined in [25]. First of all, is development of an operator behaviour model more comprehensive and realistic; it can be used not only for nuclear power plants but also for more general applications. This is a subject of current research effort by the authors.

Many researchers are moving to the integration of their studies with those of other researchers to optimise HRA techniques. Some future plans include, for example, extending AHP–SLIM into other HRAs methods to exploit its performance [19]. The method proposed by De Ambroggi and Trucco for modelling and assessment of dependent performance shaping factors through analytic network process [20] is moving towards better identification of dependencies among PSFs using the simulator PROCOS or Bayesian networks.

Bayesian networks (BN) represent, in particular, an important field of study for future developments. Many experts are studying these networks with the aim of exploiting the features and properties in the techniques HRA [44,45]. Bayesian methods are appealing since they can combine prior assumptions of human error probability (i.e. based on expert judgement) with available human performance data. Some results already show that the combination of the model conceptual causal model with a BN approach can not only qualitatively model the causal relationships between organisational factors and human reliability but can also quantitatively measure human operational reliability, identifying the most likely root causes or prioritisation of root causes of human error [44]. This is a subject of current research effort by the authors of the IDAC model as an alternative way for calculating branch probability and representing PIF states as opposed to the current method; in the current method, branch probabilities are dependent on the branch scores that are calculated based on explicit equations reflecting the causal model built, based on the influence of PIFs and other rules of behaviour.

Additional research and efforts are related to the performance shaping factors (PSFs). Currently, there are more than a dozen HRA methods that use PIFs/PSFs, but there is no standard set of PIFs used among methods. The performance shaping factors at present are not defined specifically enough to ensure consistent interpretation of similar PIFs across methods. There are few rules governing the creation, definition, and usage of PIF sets. Within the HRA community, there is a widely acknowledged need for an improved HRA method with a more robust scientific basis. Currently, there are several international efforts to collect human performance data that can be used to improve HRA [46].

Of course, many studies that are being carried out are aimed at improving the application of HRA methods in complex environments, such as nuclear power plants. The methods already developed in these areas are adapting to different situations by expanding their scope.

Author details

Valentina Di Pasquale, Raffaele Iannone*, Salvatore Miranda and Stefano Riemma

Dept. of Industrial Engineering – University of Salerno, Italy

References

[1] Iannone, R., Miranda, S., Riemma S.: Proposta di un modello simulativo per la determinazione automatica delle pause di lavoro in attività manifatturiere a prevalente contenuto manuale. Treviso - Italy ANIMP Servizi Srl Pag. 46–60 (2004).

[2] Madonna, M., et al.: Il fattore umano nella valutazione dei rischi: confronto metodologico fra le tecniche per l'analisi dell'affidabilità umana. Prevenzione oggi. 5 (n. 1/2), 67–83 (2009).

[3] Griffith, C.D., Mahadevan, S.: Inclusion of fatigue effects in human reliability analysis. Reliability Engineering & System Safety, 96 (11), 1437–1447 (2011).

[4] Hollnagel, E.: Cognitive Reliability and Error Analysis Method CREAM (1998).

[5] Hollnagel, E.: Reliability analysis and operator modelling. Reliability Engineering and System Safety, 52, 327–337 (1996).

[6] Bye, A., Hollnagel, E., Brendeford, T.S.: Human-machine function allocation: a functional modelling approach. Reliability Engineering and System Safety, 64 (2), 291–300 (1999).

[7] Cacciabue, P.C.: Modelling and simulation of human behaviour for safety analysis and control of complex systems. Safety Science, 28 (2), 97–110 (1998).

[8] Kim, M.C., Seong, P.H., Hollnagel, E.: A probabilistic approach for determining the control mode in CREAM. Reliability Engineering and System Safety, 91 (2), 191–199 (2006).

[9] Kim, I.S.: Human reliability analysis in the man–machine interface design review. Annals of Nuclear Energy, 28, 1069–1081 (2001).

[10] Sträter, O., Dang, V., Kaufer, B., Daniels, A.: On the way to assess errors of commission. *Reliability Engineering and System Safety*, 83 (2), 129–138 (2004).

[11] Boring, R.L., Blackman, H.S.: The origins of the SPAR-H method's performance shaping factor multipliers. In: Joint 8th IEEE HFPP/13th HPRCT (2007).

[12] Blackman, H.S., Gertman, D.I., Boring, R.L.: Human error quantification using performance shaping factors in the SPAR-H method. In: 52nd Annual Meeting of the Human Factors and Ergonomics Society (2008).

[13] Kirwan, B.: The validation of three human reliability quantification techniques – THERP, HEART and JHEDI: Part 1 – Technique descriptions and validation issues. *Applied Ergonomics*, 27 (6), 359–373 (1996).

[14] Kirwan, B.: The validation of three human reliability quantification techniques – THERP, HEART and JHEDI – Part 2 – Results of validation exercise. *Applied Ergonomics*, 28 (1), 17–25 (1997).

[15] Kirwan, B.: The validation of three human reliability quantification techniques – THERP, HEART and JHEDI – Part 3 – Practical aspects of the usage of the techniques. *Applied Ergonomics*, 28 (1), 27–39 (1997).

[16] Kim, J.W., Jung, W.: A taxonomy of performance influencing factors for human reliability analysis of emergency tasks. *Journal of Loss Prevention in the Process Industries*, 16, 479–495 (2003).

[17] Hallbert, B.P., Gertmann, D.I.: Using Information from operating experience to inform human reliability analysis .In: International Conference On Probabilistic Safety Assessment and Management (2004).

[18] Lee, S.W., Kim, R., Ha, J.S., Seong, P.H.: Development of a qualitative evaluation framework for performance shaping factors (PSFs) in advanced MCR HRA. *Annals of Nuclear Energy*, 38 (8), 1751–1759 (2011).

[19] Park, K.S., Lee, J.: A new method for estimating human error probabilities: AHP–SLIM. *Reliability Engineering and System Safety*, 93 (4), 578–587 (2008).

[20] De Ambroggi, M., Trucco, P.: Modelling and assessment of dependent performance shaping factors through analytic network process. *Reliability Engineering & System Safety*, 96 (7), 849–860 (2011).

[21] Trucco, P., Leva, M.C.: A probabilistic cognitive simulator for HRA studies (PROCOS). *Reliability Engineering and System Safety*, 92 (8), 1117–1130 (2007).

[22] Leva, M.C., et al.: Quantitative analysis of ATM safety issues using retrospective accident data: the dynamic risk modelling project. *Safety Science*, 47, 250–264 (2009).

[23] Boring, R.L.: Dynamic human reliability analysis: benefits and challenges of simulating human performance. In: Proceedings of the European Safety and Reliability Conference (ESREL 2007) (2007).

[24] Boring, R.L.: Modelling human reliability analysis using MIDAS.In: International Workshop on Future Control Station Designs and Human Performance Issues in Nuclear Power Plants (2006).

[25] Mosleh, A., Chang, Y.H.: Model-based human reliability analysis: prospects and requirements. *Reliability Engineering and System Safety*, 83 (2), 241–253 (2004).

[26] Mosleh, A., Chang, Y.H.: Cognitive modelling and dynamic probabilistic simulation of operating crew response to complex system accidents – Part 1: Overview of the IDAC model. *Reliability Engineering and System Safety*, 92, 997–1013 (2007).

[27] Mosleh, A., Chang, Y.H.: Cognitive modelling and dynamic probabilistic simulation of operating crew response to complex system accidents – Part 2: IDAC performance influencing factors model. *Reliability Engineering and System Safety*, 92, 1014–1040 (2007).

[28] Mosleh, A., Chang, Y.H.: Cognitive modelling and dynamic probabilistic simulation of operating crew response to complex system accidents – Part 3: IDAC operator response model. *Reliability Engineering and System Safety*, 92, 1041–1060 (2007).

[29] Mosleh, A., Chang, Y.H.: Cognitive modelling and dynamic probabilistic simulation of operating crew response to complex system accidents – Part 4: IDAC causal model of operator problem-solving response. reliability engineering and system safety, 92, 1061–1075 (2007).

[30] Mosleh, A., Chang, Y.H.: Cognitive modelling and dynamic probabilistic simulation of operating crew response to complex system accidents – Part 5: Dynamic probabilistic simulation of the IDAC model. *Reliability Engineering and System Safety*, 92, 1076–1101 (2007).

[31] http://www.hse.gov.uk/research/rrpdf/rr679.pdf

[32] http://www.cahr.de/cahr/Human%20Reliability.PDF

[33] http://conference.ing.unipi.it/vgr2006/archivio/Archivio/pdf/063-Tucci-Giagnoni-Cappelli-MossaVerre.PDF

[34] http://www.nrc.gov/reading-rm/doc-collections/nuregs/contract/cr6883/cr6883.pdf

[35] Jansen, N.W.H., Kant, I., Van den Brandt, P.A.: Need for recovery in the working population: description and associations with fatigue and psychological distress. *International Journal of Behavioral Medicine*, 9 (4), 322–340 (2002).

[36] Jett, Q.R., George, J.M.: Work interrupted: a closer look at the role of interruptions in organizational life. *Academy of Management Review*, 28 (3), 494–507 (2003).

[37] Tucker, P., Folkard, S., Macdonald, I.: Rest breaks and accident risk. *Lancet*, 361, 680 (2003).

[38] Folkard, S., Tucker, P.: Shift work, safety, and productivity. *Occupational Medicine*, 53, 95–101 (2003).

[39] Folkard, S., Lombardi, D.A.: Modelling the impact of the components of long work hours on injuries and "accidents". *American Journal of Industrial Medicine*, 49, 953–963 (2006).

[40] McLean, L., Tingley, M., Scott, R.N, Rickards, J.: Computer terminal work and the benefit of microbreaks. *Applied Ergonomics*, 32, 225–237 (2001).

[41] Van Den Heuvel, S.G., et al.: Effects of software programs stimulating regular breaks and exercises on work-related neck and upper-limb disorders. *Scandinavian Journal of Work, Environment & Health*, 29 (2), 106–116 (2003).

[42] Jaber, M.Y., Bonney, M.: Production breaks and the learning curve: the forgetting phenomenon. *Applied Mathematics Modelling*, 20, 162–169 (1996).

[43] Jaber, M.Y., Bonney, M.: A comparative study of learning curves with forgetting. *Applied Mathematics Modelling*, 21, 523–531 (1997).

[44] Li Peng-cheng, Chen Guo-hua, Dai Li-cao, Zhang Li: A fuzzy Bayesian network approach to improve the quantification of organizational influences in HRA frameworks. *Safety Science*, 50, 1569–1583 (2012).

[45] Kelly, D.L., Boring, R.L., Mosleh, A., Smidts, C.: Science-based simulation model of human performance for human reliability analysis. Enlarged Halden Program Group Meeting, October 2011.

[46] Groth, K.M., Mosleh, A.: A data-informed PIF hierarchy for model-based human reliability analysis. *Reliability Engineering and System Safety*, 108, 154–174 (2012).

Permissions

The contributors of this book come from diverse backgrounds, making this book a truly international effort. This book will bring forth new frontiers with its revolutionizing research information and detailed analysis of the nascent developments around the world.

We would like to thank Massimiliano Schiraldi, for lending his expertise to make the book truly unique. He has played a crucial role in the development of this book. Without his invaluable contribution this book wouldn't have been possible. He has made vital efforts to compile up to date information on the varied aspects of this subject to make this book a valuable addition to the collection of many professionals and students.

This book was conceptualized with the vision of imparting up-to-date information and advanced data in this field. To ensure the same, a matchless editorial board was set up. Every individual on the board went through rigorous rounds of assessment to prove their worth. After which they invested a large part of their time researching and compiling the most relevant data for our readers. Conferences and sessions were held from time to time between the editorial board and the contributing authors to present the data in the most comprehensible form. The editorial team has worked tirelessly to provide valuable and valid information to help people across the globe.

Every chapter published in this book has been scrutinized by our experts. Their significance has been extensively debated. The topics covered herein carry significant findings which will fuel the growth of the discipline. They may even be implemented as practical applications or may be referred to as a beginning point for another development. Chapters in this book were first published by InTech; hereby published with permission under the Creative Commons Attribution License or equivalent.

The editorial board has been involved in producing this book since its inception. They have spent rigorous hours researching and exploring the diverse topics which have resulted in the successful publishing of this book. They have passed on their knowledge of decades through this book. To expedite this challenging task, the publisher supported the team at every step. A small team of assistant editors was also appointed to further simplify the editing procedure and attain best results for the readers.

Our editorial team has been hand-picked from every corner of the world. Their multi-ethnicity adds dynamic inputs to the discussions which result in innovative outcomes. These outcomes are then further discussed with the researchers and contributors who give their valuable feedback and opinion regarding the same. The feedback is then collaborated with the researches and they are edited in a comprehensive manner to aid the understanding of the subject.

Apart from the editorial board, the designing team has also invested a significant amount of their time in understanding the subject and creating the most relevant covers. They scrutinized every image to scout for the most suitable representation of the subject and create an appropriate cover for the book.

The publishing team has been involved in this book since its early stages. They were actively engaged in every process, be it collecting the data, connecting with the contributors or procuring relevant information. The team has been an ardent support to the editorial, designing and production team. Their endless efforts to recruit the best for this project, has resulted in the accomplishment of this book. They are a veteran in the field of academics and their pool of knowledge is as vast as their experience in printing. Their expertise and guidance has proved useful at every step. Their uncompromising quality standards have made this book an exceptional effort. Their encouragement from time to time has been an inspiration for everyone.

The publisher and the editorial board hope that this book will prove to be a valuable piece of knowledge for researchers, students, practitioners and scholars across the globe.

List of Contributors

Fabio De Felice and Antonella Petrillo
University of Cassino, Department of Civil and Mechanical Engineering, Cassino, Italy

Stanislao Monfreda
Fiat Group Automobiles EMEA WCM Cassino Plant Coordinator, Cassino, Italy

Vittorio Cesarotti and Vito Introna
University of Rome "Tor Vergata", Italy

Alessio Giuiusa
University of Rome "Tor Vergata", Italy
Area Manager Inbound Operations at Amazon.com

Raffaele Iannone
Department of Industrial Engineering, University of Salerno, Italy

Maria Elena Nenni
Department of Industrial Engineering, University of Naples Federico II, Italy

Filippo De Carlo
Industrial Engineering Department, University of Florence, Florence, Italy

Marcello Fera, Alfredo Lambiase and Giada Martino
University of Salerno - Dpt. of Industrial Engineering, Fisciano (Salerno), Italy

Fabio Fruggiero
University of Basilicata - School of Engineering, Potenza, Italy

Maria Elena Nenni
University of Naples Federico II - Dpt. of Economic Management, Napoli, Italy

Francesco Giordano and Massimiliano M. Schiraldi
Department of Enterprise Engineering, "Tor Vergata" University of Rome, Italy

Alberto Regattieri
DIN - Department of Industrial Engineering, University of Bologna, Bologna, Italy

Giulia Santarelli
DTG - Department of Management and Engineering, University of Padova, Padova, Italy

Giulio Di Gravio, Francesco Costantino and Massimo Tronci
Department of Mechanical and Aerospace Engineering, University of Rome "La Sapienza", Italy

Valentina Di Pasquale, Raffaele Iannone, Salvatore Miranda and Stefano Riemma
Dept. of Industrial Engineering – University of Salerno, Italy

Printed in the USA
CPSIA information can be obtained
at www.ICGtesting.com
JSHW011431221024
72173JS00004B/752